Vineyards, Rocks, and Soils

The Wine Lover's Guide to Geology

ALEX MALTMAN

OXFORD
UNIVERSITY PRESS

OXFORD
UNIVERSITY PRESS

Oxford University Press is a department of the University of Oxford. It furthers
the University's objective of excellence in research, scholarship, and education
by publishing worldwide. Oxford is a registered trade mark of Oxford University
Press in the UK and certain other countries.

Published in the United States of America by Oxford University Press
198 Madison Avenue, New York, NY 10016, United States of America.

© Oxford University Press 2018

Library of Congress Cataloging-in-Publication Data
Names: Maltman, Alex, 1944– author.
Title: Vineyards, rocks, and soils : the wine lover's guide
to geology / Alex Maltman.
Other titles: Wine lover's guide to geology
Description: New York, NY : Oxford University Press, [2018] | Includes index.
Identifiers: LCCN 2017048404 | ISBN 9780190863289
Subjects: LCSH: Wine and wine making—Chemistry. |
Soil chemistry. | Geochemistry. | Geology.
Classification: LCC TP548 .M31737 2018 | DDC 663/.2—dc23
LC record available at https://lccn.loc.gov/2017048404

19 18 17 16 15 14

Printed by Integrated Books International, United States of America

CONTENTS

FOREWORD

BY ANDREW JEFFORD

Wine fascinates. It's an intoxicant with a sensual identity of sumptuous diversity: a doubly heady combination. We know nothing else like it. Discovery is usually followed by enthusiasm (and overindulgence). Then, later, a question: Why? Why does this wine taste like this, and that wine like that? Why does this scent and taste mean a $1000 bottle, and that a $10 bottle? Why should this wine slumber out two decades before being broached, and that bottle be best emptied the week after next?

The simplest answer is to assign the differences between wines to four factors: grape variety, crop load, winemaking, and season. They explain much, but not everything. A select band of grape varieties is now grown the world over, and viticultural and winemaking knowledge is so freely shared that it's barely an exaggeration to say that most of the world's ambitious wines are cropped and made in a similar manner. The results—between continents, countries, regions, and sometime even between vineyards themselves, during the same season, from the same cellar—remain sumptuously diverse. There is a thistly residuum to wine's character that the customary inputs do not seem to be able to account for. We make our way back to the original question.

The instinctive response is to look down to the ever-changing medium in which the vine is rooted and to ascribe the differences to the soil. The soil is visible, palpable, scrutable. It can be dug into, and its layers can be revealed and photographed. We can name the rock(s) that lie(s) beneath. Comparisons can easily be drawn. We understand that plants possess roots; we assume—since we cannot see vines drink the air—that it is from their roots that vines draw their nutrition. We note that vines root deeply and that they grow happily in stony, dry places. We sip delicious wines with the images of such places in our minds. We confuse circumstance and cause.

From our simple yet fallible observations comes an entire interpretative tradition in wine writing and wine scholarship. Scents and flavors are assigned mineral triggers. Tasteless minerals, indeed, acquire aromas and flavors of their own.

Appellation boundaries are based on rock types. Geological terminology litters wine labels, sprinkles wine textbooks, and comes tumbling from the lips of somme-liers. Received truths about the necessary impacts of "limestone soils" or "granite soils" are unthinkingly recycled. Wine fairytales, if they come sprinkled with a little geological terminology, emerge with the foot-stomping force of a gospel truth choir.

Alex Maltman's book is an essential call to order. With patient clarity, he explains the geological fundamentals that every wine lover should know (in addition to its other virtues, this book is a limpid introduction to basic geological concepts). He explains what a vine is, where it finds the source of its sustenance, and what it can and cannot do with soil minerals. He points out many of the geological errors, mis-conceptions, and misunderstandings with which wine description and literature is littered. He offers the wine world a terminological reset. The chance to start anew, our language cleansed and calibrated, makes this book essential reading for anyone who wishes to teach or write about wine or even think clearly about wine.

There are a couple of things, though, that Alex Maltman doesn't do. The first is that he does not attempt to ban metaphor. Wine descriptions are a seething meta-phorical mass, indisciplined and subjective, and if those writing descriptions wish to allude to our sensual experiences of looking at, walking on, or working with earth or rock, then they are welcome to do so (though Maltman does point out that the sensual print of the soil, for which the term *petrichor* was coined in 1968, is based not on its mineral content but on its organic matter). He simply stresses that we should not take wine metaphors concerning flint or slate any more literally than those invoking raspberries or cream.

The second thing he doesn't do is to suggest that soils, and the rock types on which they are in part based, are insignificant or play no role whatsoever in creating wine aroma and flavor. The book's conclusion is an acknowledgment both of the extraordinary complexity of inputs that go into creating aroma and flavor and of the manifold gaps in our understanding of these processes. He is not an iconoclast for the sadistic fun of it; he is a scientist—and wine lover—with an open and enquiring mind who merely asks that we should understand what technical terms mean before we use them and that we should respect the journey toward understanding which science has so far permitted us.

What he certainly does do, however, is acknowledge, indeed share, the sense of wonder about the relationship between wine and place that most wine lovers come to feel, something that this book will deepen as well as clarify and set limits to.

Geology is planetary history. Together with cosmology, it offers human beings the largest perspectives to which we can aspire. These vast vistas and abyssal pros-pects are a constant challenge to the imagination. They set us and our works (of which wine creation is not the least) in humbling context; they inspire, shock, invigorate. You cannot complete the human experience without making room for geological insight. In addition to its other virtues, this book will help wine lovers on that path.

PREFACE

A bond between wine and the land has long been cherished as something special, and in recent years the idea has reached new heights. In particular, today's wine writers like to enthuse about vineyard geology: the rocks and soils in which the vines grow. Wineries proclaim the uniqueness of their vineyards' geology, and promoters have latched on to the "sense of place" this implies. Commentators boldly assert the special qualities this or that bedrock brings to wine; plenty of enthusiasts are convinced that the vineyard geology can actually be tasted in their wine glass. Thus, writings about Burgundy's Côte d'Or rhapsodize about *smectite* and *colluvium*; numerous wines are named after types of rock—*orthogneiss, amphibolite*, and *mylonite*, among them—so that these days it's almost obligatory in the wine world to know something about the geology of wine-producing areas and of particular vineyards. As one blogger put it, "nowadays wine lovers need to know their *tuff* from their *tufa*, and their *slate* from their *schist*."

Unfortunately, vineyard geology is commonly misunderstood. I often see terms being misused, processes confused, and explanations misrepresented. But then geology is a highly conceptual subject and not easy to pick up quickly. Most geological processes are unfamiliar. They happen imperceptibly slowly, often underground, out of sight; many took place unimaginably long ago. Geological things tend to be hard to classify: even rocks don't readily fall into separate divisions. And because geological features are enmeshed in the Earth's complex interlinked systems, they are hard to grasp from an isolated, terse definition.

Moreover, because modern geology grew rapidly in several places at the same time, different words were simultaneously coined for the same things, even within a single country. Today, international panels are struggling to clarify this legacy and to demarcate some uniformity of usage. As just one example, the committee that was convened to look at igneous rocks found as it began its work that well over a thousand different names were in use. It's hardly surprising, then, that vineyard commentaries can be confused about the geology!

At the same time, there is no ready explanation of geology that the wine lover can turn to, in the context of vineyards and wine. This book aims to provide just that explanation. True, there are plenty of books *describing* the geology of particular wine-producing areas and countries, and there are glossaries around—two or three lines tersely defining (rather than explaining) a term (and, it has to be said, some with instances that are, to this professional geologist, just plain wrong). Web pages vary in authority and scope: search engines are manifestly not solving the problem.

Therefore, in this book I introduce and explain geology from first principles, and throughout from the perspective of vineyards and wine. Key meanings of terms within the explanations are in bold font, and all are indexed. My hope is that as a result any wine lover can make a quick check on some relevant geological matter, while the serious enthusiast can read at greater length to get a coherent understanding of geology as it applies to vineyards and wine. Also, almost all of the writings that describe the geology of wine regions make little attempt, apart from a vague remark or two about drainage, nutrition, or soil warmth, to say *why* the geology may be relevant, let alone evaluate its importance. So I also tackle these matters here, particularly in the contexts of minerality, terroir, and wine flavor.

This book, then, introduces basic geology in the context of wine. (Incidentally, the word "geology" originally simply meant the scientific study of the solid Earth, but these days it has expanded to encompass all the materials and processes that are involved. And in general I've used the form "geological" for mental constructs—time, for example—and the American convention "geologic" for more tangible things.) In contrast with standard primers on geology, there are no chapters here on the Earth's interior, fossils, volcanoes, hydrocarbons, magnetism, glaciers, meteorites, and so on. Fascinating though these topics are, they appear at best to be distant to the world of wine. Rather, my emphasis is on the kinds of processes that can shape vineyards, as well as on the minerals, rocks, and soils that host the vines. The first chapter provides a preview. I have suggested further readings at the end of some chapters but have restricted them to introductory items that are reasonably accessible or easily obtained. There are other boundaries to the book's scope. It's not about the practicalities of managing vineyard soils, nor is it primarily about grapevines and wines. We tend to forget that virtually all of the world's vineyards are no longer in their initial, natural state. Indeed, many are routinely manipulated with sophisticated systems of drainage, fertilization, irrigation, and the rest. None of these are discussed here. The book is concerned with explaining the starting point, the natural baseline geology, which a grower may or may not be choosing to modify.

I have tried to keep terminology to a minimum, presenting terms I see being used in the wine world. I'm well aware that informed readers may question some of the definitions I use—of schist, for instance, weathering, or escarpment, or even of soil—but I feel it would be hopelessly confusing to mention all the different nuances that are out there. The compass of subjects I cover is also a personal one, guided by what I come across most often in wine writings. So if a reader looks in vain for an explanation of some geological word that has caused puzzlement, I can only apologize: it's not possible for the coverage to be comprehensive. Incidentally, by "wine writings," a phrase I use frequently in the book, I mean wine labels, trade promotional materials, magazines, newspaper columns, books, and, of course, today's plethora of wine blogs.

The usage of geological terms is constantly evolving, with new terms being introduced and old favorites discarded, so I have had to make some slightly arbitrary but pragmatic decisions, based on how I see things trending in wine writings. For example, I see no sign that some time-honored names for igneous rocks are disappearing, and so I have continued to use them here, even if they are out of line with modern technical usage. But my treatment of terms for geological time is the opposite. Some of the international terms for geological time are increasingly being adopted, such that their parochial forerunners are already looking decidedly obsolete. No doubt if I were writing this book in another twenty years, some of my judgments on appropriate terminology would be different.

I have written the book in a casual teaching style that seems fitting; I apologize if for some readers this informality grates. In places, to give a human face to geology, I have recalled the story behind a term or remembered the pioneer who first saw the light. Even so, some passages will probably strike some readers as dauntingly technical—such as those explaining ions, silicate tetrahedra, and pH—so I've separated those into text boxes. Thus, they can easily be skipped, but I do think they're needed for a proper understanding of how vineyard geology works. Conversely, I apologize if my attempts to give clear explanations lead some readers to think them a bit labored. But decades of teaching undergraduate students have taught me that is a small price to pay if others are reaching a new clarity of understanding.

At one level, using geological words nonchalantly—say, calling any hard rock a granite—is harmless enough. But if we are to comprehend how vineyards work, then we have to be more systematic, more scientific. Of course, ultimately, our appreciation, our sensual response to wine, is beyond scientific analysis, and I know there are those who feel that the scientific approach has already interfered too much in wine and is destroying its magic. But I think the opposite. For me, understanding vineyard geology enriches the world of wine. The great English landscape painter, John Constable, studied the land and the sky in order to inform his painting. "We see nothing truly until we understand it," he said, and the same could be said of looking at vineyards. Besides improving the all-round standard of wine, science fine tunes our discernment of what's going on out there, and ultimately it enhances our appreciation. It continues to reveal the astonishing intricacies and dazzling beauty that are involved when vines grow and grape juice becomes wine.

ACKNOWLEDGMENTS

I've always been struck by the dedication and enthusiasm of wine people around the world and by their generosity. Individuals far too numerous to name have willingly helped me on my way. Growers have always found time to show me their vineyards, and back in the winery so often my probing questioning of winemakers has turned into sessions of sitting around tables of half-emptied bottles and scurries to get yet more samples. I thank all of you.

Andrew Jefford's eclectic and insightful writings are an inspiration to me, and so I was flattered when he agreed to write a Foreword to this book. Thanks, Andrew, for your support and encouragement. Professor Mike Summerfield, one of the world's leading geomorphologists, kindly read and commented on the chapter on landforms, an area where I have less first-hand experience. Some of the book's passages, especially in Chapters 6 and 11, have already appeared in the *World of Fine Wine* magazine. I thank the magazine's editor, Neil Beckett, for allowing me to use them here.

My wife, Joanne, read through the entire text: she has always supported my efforts to try and find out how wine works. Once upon a time I dragged her and my two little children around wineries when they would much rather have been playing on the beach. Today their mother unflinchingly continues the tradition, trailing with me to obscure places just because some vines are growing there.

ABBREVIATIONS AND CONVERSIONS

Throughout the book, I refer to cation exchange capacity (explained in Chapter 2) as CEC. PV is precision viticulture.

I give the amounts for the nutrients in wine in ppm—parts per million, which is equivalent to concentrations in milligrams per liter. Ppb and ppt are parts per billion and parts per trillion, respectively.

For some countries, I refer in places to their familiar appellation systems for wine regions. Thus, France's AOC means *Appellation d'Origine Contrôlée,* and IGP is the *Indication Géographique Protégée* designation, which is replacing the previous *Vins de Pays.* Italy's DOC stands for *Denominazione di Origine Controllata,* and DOCG for *Denominazione di Origine Controllata e Garantita.* Spain's DO means *Denominación de Origen.* In Austria, DAC stands for *Districtus Austriae Controllatus.* AVA is American Viticultural Area, the United States system, and VQA is Vintners Quality Alliance, the Canadian equivalent.

All the maps have north toward the top of the page.

Chemical symbols used in the book are as follows:

Al   aluminum

Ca   calcium

C   carbon

Cl   chlorine

H   hydrogen

Fe   iron

N   nitrogen

O   oxygen

P   phosphorus

K   potassium

Si   silicon

Na   sodium

S   sulfur

A centimeter is 10 millimeters, approximately equivalent to 0.4 of an inch. One inch is 2.54 centimeters or 25.4 millimeters. A kilometer is approximately 0.6 of a mile.

One metric ton, or tonne, is equivalent to 1.1 United States tons or 0.98 Imperial tons. One hectare is 2.47 acres.

All temperatures are given in degrees Celsius, °C, the convention in science and most of the world. Regarding values mentioned in the text, 25°C is equivalent to 77° Fahrenheit; 100°C is 212°Fahrenheit; 200°C is 392° Fahrenheit; 300°C is 572° Fahrenheit; and 600°C is 1112° Fahrenheit.

1

What Are Vineyards Made Of?

What strikes you first when looking at a vineyard? Perhaps the vines themselves? Your eye may be caught by random scatterings of gnarly old bushes or by the military neatness of rows of trained vines, luxuriant in foliage in summer and little more than gaunt woody skeletons in winter. But possibly more striking might be the land itself—the geology, or at least manifestations of it. The vines may extend across a vast, flat plain, or they may be perched on a vertiginous slope, or anywhere in between—it depends on the bedrock geology. How well the vines grow will be influenced by how that bedrock weathers into soil and how the vine roots respond. The soil may have an eye-catching color or may be astonishingly stony, consisting of little more than rock debris. This quality, too, depends on the geology.

But what exactly is this vineyard ground? What are such things as bedrock, soil, and stone made of? Where do they come from? How did they get this way? The answers form the basis of understanding vineyard geology, so let's begin here, with a few fundamentals.

A Glimpse of the Very Basics: Elements, Atoms, and Ions

We can think about what the ground in a vineyard is made of in three ways. *The first way* is that, like all matter, it consists of atoms of chemical elements. And remarkably, although there are nearly a hundred different chemical elements in nature, the ground is dominated by just eight of them (Figure 1.1a). You could even say that it's pretty much made up of only four of these elements, as the first four on the list account for nearly 88% of the composition. Preponderant among them are oxygen, at no less than 46%, and silicon, at 28%. So there's a lot of these two elements in most vineyards!

As an aside, it's the same kind of story with living organisms: about 95% of their composition consists of just three elements—carbon, oxygen, and hydrogen—and that includes grapevines and grapes (Figure 1.1b). This fact is a reminder that over 99% of wine is just carbon, oxygen, and hydrogen! We call matter that is made

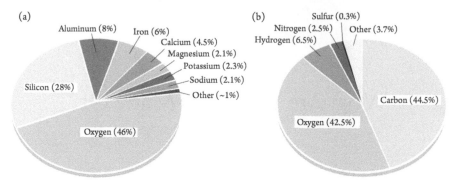

Figure 1.1 Vineyards in terms of chemical elements. (a) Soil geology, as an average chemical composition. (b) The average chemical composition of biological organisms, including vines and grapes. Inorganic constituents of wine typically total less than 0.4%, and almost half of this is potassium.

primarily of these elements **organic**, as opposed to **inorganic** substances, most of which lack carbon and hydrogen but involve metals. And, incidentally, that's what is meant by "organic" in this book, and not organic methods of viticulture.

So we can view the ground as elements, but this has to be just part of the story. Look at all that oxygen, for example. We tend to think of oxygen as a gas, as in the air around us, but clearly that can't be the situation here: vineyards don't bubble! In reality, most elements exist in nature not as independent elements but as elements combined into **compounds**. The outer part of an atom is clouded with tiny, negatively charged particles called electrons, which tend to intermix with electrons in other elements; the result is an interlinked mass, a chemical compound. Such atoms, whose number of electrons has changed, are known as **ions**. Losing electrons leaves an atom with a positive electrical charge, which we call a **cation**; the converse is an **anion**. This whole idea of atoms taking on a charge, and indeed the term *ion* itself, came about through experiments on electricity conducted a couple of centuries ago by the great English scientist Michael Faraday (1791–1861). Presumably, this self-educated blacksmith's son from south London never imagined that his discovery would relate to grapevines and wine! But it certainly does: among other things, ions are central to the nutrition of vines. We'll be seeing quite a lot of them in the chapters to come.

What Exactly Are Ions?

Atoms are not the indivisible little balls once imagined, but themselves have constituent particles, lots of them, some of which are almost the stuff of science fiction. Only one need concern us here: the **electron**. The name comes from the Latin word *electrum* for amber, a natural resin with the ability to attract small objects when rubbed. The great wordsmith Sir Thomas Browne (originator of a cornucopia of words, such as cryptography, coma, compensate, and computer) called the effect "electricity". We now know that the attraction—and electricity itself—comes about through electrons, with each particle carrying a negative electrical charge. These tiny particles cloud the outer parts of an atom, in numbers depending on the particular element and, because they are more or less loosely held, they control how the element behaves in nature.

A few elements have the perfect complement of electrons for their atoms to exist happily in nature as stable entities, independent from each other. Gold is the greatest example of this independence. Gold doesn't want to combine with anything else; in fact, its inertness is the basis of its legion of uses. After all, we would hardly prize gold for jewelry if it reacted with water or chemicals in the air! However, the atoms of most elements are not inert in this way; rather, they have either too many or too few electrons for them to exist stably by themselves. If an element that would profit from losing or sharing electrons can associate with another element that would benefit from gaining electrons, then together they could form a stable **compound**. In some situations, say with sodium chloride, the sodium donates an electron to the receiving chlorine (Figure 1.2); in

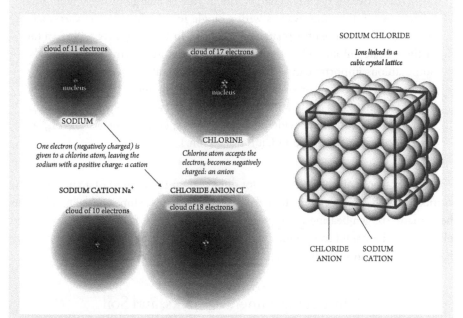

Figure 1.2 Diagram showing a sodium atom giving one electron (hence becoming a cation) to a chlorine atom (which therefore becomes an anion) to bond the two elements into the compound called sodium chloride (the mineral halite, common salt). The substance is crystalline because its ions are arranged systematically (see Chapter 2), which in the case of salt is a cubic pattern.

others, such as silicon dioxide—the mineral quartz—the electrons are shared between the silicon and the oxygen.

In either case, each constituent atom in a compound, now effectively having either fewer or more of these electrons than originally, will have acquired an electrical charge, and so the atom is properly called an **ion**. Losing one electron, with its single negative charge, leaves the ion with a single positive charge, and we now call it a **cation** (pronounced cat-eye-on, with the accent on cat). This is indicated in its chemical symbol by adding a single plus sign, as in potassium, K^+. Elements that lose two electrons, for example, magnesium, are written Mg^{2+}. We add more plus signs as further electrons are lost, such as in the aluminum ion, Al^{3+}. Those elements that gain electrons, thus becoming negatively charged, are called **anions**. Negative signs are added to their symbol to indicate the degree of charge, that is, the number of electrons gained. Cl^-, chloride, represents a single charge, with one electron added; O^{2-}, oxide, shows that the oxygen atom has gained two electrons. (Both of these examples can, of course, also exist in uncharged form, as the elements chlorine and oxygen.)

In some cases, the anion itself can consist of two elements linked together but with an overall negative charge. The degree of charge of the linked ions is indicated in the same way as with a single ion, such as in the carbonate anion, CO_3^{2-}, and the phosphate anion, PO_4^{3-}. In Chapter 3 we will see a lot of the silicate anion, SiO_4^{4-} because it underpins many vineyard materials. Some anions important in wine can involve several elements, such as tartrate, $C_4H_4O_6^{2-}$. However, only one cation comprises more than one element linked together. This is the ammonium ion, NH_4^+, and it is important in the vineyard by helping to provide nitrogen to vines. (Ammonia gas is NH_3, so, being an uncharged compound, it is not an ion.)

In other words, although there's a lot of oxygen in the ground of a vineyard, it is combined with other elements to make solid compounds. More precisely, vineyards have many oxygen-bearing compounds. Such substances, these rigidly connected elements—these solid compounds within the ground—are called *minerals*. Thus, we have *the second way* of thinking about what composes vineyards: geologic minerals.

Introducing Minerals, Rocks, and Soil

Often it's more useful to think not of the chemical elements that constitute vineyard ground but of the compounds they make, the **minerals**. When we say, for example, that the vineyards at Cornas, France, have lots of mica or that quartz is noticeable around Hawke's Bay, New Zealand, we are referring to minerals. Clearly, then, minerals are highly relevant to vineyards, and so we look at them closely in the next two chapters. But let's be clear about one thing from the outset, inasmuch as in the wine world the word "mineral" leads to confusion and misunderstanding (Chapter 9). The present discussion is about minerals in the geological sense, compounds such as gypsum, calcite, and feldspar.

There are thousands of minerals, but here's a remarkable fact about them: In nature we find again and again the same few of them collected together in roughly similar proportions, and themselves bonded together to form strong, solid masses. The name for these rigid collections? Rock. So this is *the third way* of looking at the ground of a vineyard: it's made of rock (together with, as we shall see shortly, detritus derived from it that helps make soil). A **rock** is a coherent, rigid, solid aggregate of minerals. The constituent minerals do not share electrons like the ions within compounds; rather, they're bonded together in several ways, which we'll cover in future chapters. In view of what some of the literature on wine says, here's another point we should make clear: minerals combine to make rock. Thus *rocks are not the same as minerals*; the words are not synonymous.

We think of the Earth as a rocky planet; its outer rind, called the **crust**, is made of rock, as is the underlying **mantle**, and Earth's central **core**. The outermost part of this rocky crust is **bedrock**. It's everywhere beneath our feet; all vineyards are founded, at some depth or other, on bedrock. In places such as craggy hillsides or cliffs, the bedrock may be visible, protruding as **outcrops** (Figure 1.3; see Plate 1). Scattered around may be large fragments broken from the bedrock, perhaps locally derived or brought there from much further afield, perhaps by rivers. Sometimes we see in mountainous vineyards, such as the ones at Elqui (Chile) or Pritchard Hill in the Napa Valley (California), just the tops of enormous boulders. Despite their size, they are nevertheless fragments; they are loose, detached from bedrock. If struck hard enough, they will move ever so slightly, with a dull thud. Bedrock responds to striking with a sharp, metallic ring because it is literally joined on, so to speak, to the rest of the planet.

Following are two illustrations of how vineyard ground involves elements, ions, minerals, and rocks, situations which we'll be exploring further in this book. Carbon shares electrons with oxygen to form the *carbonate* anion, and where its negative charge is balanced by the cation calcium, the resulting mineral is *calcite*, the basis of **calcareous** materials. The rocks limestone and marl are calcareous, which, being the bedrock of a number of classic French vineyards, are important in the wine world and for some wine enthusiasts have particular allure. But the vast majority of rocks, whether in vineyards or anywhere else, involve minerals based on the **silicate** anion, SiO_4^{4-}, in which one silicon atom shares its electrons with four oxygen atoms. These are **siliceous** materials; granite is an example. The word "siliceous" may be less familiar than calcareous and may sound technical and modern, but it was being used as long ago as 1788, by one of geology's heroes: James Hutton (1726–1797). We'll soon be meeting this geological luminary again.

Now, let's take a preliminary look at the nature of rocks. First, they're awkward to classify and always have been. They don't fall naturally into boxes like chemical

Figure 1.3 Outcrops of bedrock around a vineyard at Cederberg Winery, Western Cape, South Africa. There are some large, loose boulders (e.g., bottom left), but most of the bare rock surfaces in the foreground are outcrops of the slate bedrock. The vines in the valley are growing in slaty soil, largely debris from the foreground rocks. On the hillside on the far side of the valley there are further outcrops of bedrock, but consisting there of stratified red sandstone, showing layers that in this view are inclined downward toward the left of the photograph (south). Otherwise, the opposite slope is covered by a sandy soil that is only supporting scrubby vegetation.

elements or minerals or, for that matter, animals. A giraffe is either a giraffe or it isn't: in general, rocks aren't like that. The rocky soils of Priorat in northeast Spain are sometimes described as slate and other times as schist. Neither description is wrong because this material falls into a "gray area" between two rock types. Early attempts at classifying rocks tried to copy Carl Linnaeus's system which had proved so successful for living things. In one scheme from 1771, for instance, metamorphic rocks were divided by color, as in *schistus viridis, schistus purpurascens*, and *schistus niger*. But this idea didn't work. The obvious approach to a classification would utilize which particular minerals have combined to form the rock. Unfortunately, however, this scheme also turns out to be wholly unworkable and unhelpful in practice. Rocks made of the same collections of minerals can look vastly different and can occur in very different circumstances.

The most useful rock classification scheme thus far devised is a very curious one indeed. It's based on how the rock formed: Was it once molten? Was it deposited as sediment? Has it undergone changes? Yet, with only a few exceptions, no one has ever witnessed such things or is ever likely to do so. Consequently, naming a rock involves a major subjective inference, and there are plenty of examples of geologists disagreeing on what to call a particular rock. Moreover, the resulting single rock name can cover a wide variety of appearances and properties.

Despite these shortcomings, this approach has never been improved upon, and it's standard today. Interestingly in the wine context, it was the originator of this genetic classification of rocks, Henri Coquand, who in 1856 touted the notion that bedrock geology influences the quality and taste of wine. He made his pronouncement with regard to the brandy produced in the area around Cognac, France. However, the maps of Cognac quality and bedrock don't correspond, as Coquand well knew. Thus, as the whole proposal was part of a playful after-dinner speech, it has been argued that he never intended the idea to be taken seriously. Little did he know.

Igneous rocks (Chapter 4) form by solidifying from a melt. The process is easy to visualize, as we have all seen images of molten lava pouring out of a volcano, cooling as it flows until it finally solidifies—to give an igneous rock. Similar kinds of processes also happen underground, out of sight. Parts of the Earth's interior are hot enough for the rocks to exist in a molten state, and with time such melts will try to move upward, reaching progressively cooler parts until the mass solidifies. With further passing of time, the congealed material may be forced upward at the same time as the ground above is being worn away, and eventually the mass may itself be exposed at the Earth's surface. Thus, igneous rock that formed deep below ground can become visible to us and can influence the landscape (Chapter 8). Granite and basalt are examples of igneous rocks. We infer their molten origin from the tight interlocking of the constituent minerals and the lack of the features characteristic of the other classes of rock, such as layering, fossils, and signs of change.

As soon as igneous rock—or indeed any other kind of rock—is exposed at the Earth's surface, it is subjected to processes of **weathering** (Chapter 9). The rock becomes progressively broken into loose particles, which we call **sediment** (Chapter 5). Then, through the actions of gravity, running water, ice, and wind, the sediment moves across the land surface, causing **erosion** (Chapter 8). If the sediment is associated with a river, we call it **alluvium** (adjective **alluvial**), and if it finds its way to the sea, we refer to it as **marine**. At any stage, and especially if the material gets as far as the sea bottom, the sediment may reside long enough for the particles to become bonded together into a coherent mass. This solid aggregate of particles is termed a **sedimentary rock**; limestone, sandstone, and shale are common examples.

Even if the material accumulates and hardens below the seafloor, just as with igneous rocks it may later be uplifted and brought to the land surface. If the rocks are exposed, we will probably see the accumulated layering of the original sediment—the in-built **stratification** or **bedding** that is the hallmark of sedimentary rocks.

Commonly, it will no longer be in its original horizontal disposition but will have been tilted. Geologists normally view this tilting as a *downward* inclination from the horizontal and call it the **angle of dip**.

Now that the rock is exposed at the land surface, it will be attacked by weathering and erosion, and so the whole process starts over again. It was James Hutton, whom we've just met, who first grasped this notion of rocks being recycled through geological time. In his monumental *"Theory of the Earth"* published in 1788, he famously expressed his vision in a couple of memorable aphorisms. On realizing the ongoing time involved in such reworking of rocks, he asserted that "we find no vestige of a beginning, no prospect of an end." Later, reflecting on the concept, Hutton wrote that his "mind seemed to grow giddy by looking so far into the abyss of time." The genius of the discovery was recalled in Repcheck's *The Man Who Discovered Time*, but for many today Hutton is the "Father of Modern Geology," a title found on his decidedly inconspicuous grave in Edinburgh's Greyfriars Churchyard.

The hardened sedimentary material below the seafloor will in some situations not be uplifted but instead will be buried further. As it finds itself deeper underground, it will be progressively weighed down by sediment accumulating above it and warmed with the Earth's internal heat. Various physical rearrangements and redistributions of the ions forming the minerals can occur, changes that are called **metamorphism** and that result in a **metamorphic rock** (Chapter 6). At any stage, the metamorphic changes can be arrested, and, as with the other rock types, the material can be exhumed. Slate, schist, gneiss, and marble are examples of metamorphic rocks. Substantial regions of the Earth's land surface are made of metamorphic rocks, including numerous vineyard areas, such as parts of Stellenbosch (South Africa), Central Otago (New Zealand), and the Adelaide Hills (Australia).

These then are the three divisions of rocks—igneous, sedimentary, and metamorphic (though once a wayward student answered an exam question on the subject by declaring that the three main kinds of rock are classic, punk, and hard!). But let's return to the loose sediment that is produced by weathering. If it resides at the land surface for any length of time, there will be a certain amount of air and water between the particles, and rotted biological material, **humus**, may become mixed in. The material will now be capable of allowing plants to grow in it, and thus we have **soil**.

The word "soil" has various usages, but I define it as the loose sediment on the land surface in which plants can grow. It is the moisture and humus that make it soil, as both are required by plants (Chapter 9). The Moon is covered in rock debris, but it lacks water and humus, and so it has no soil (even though many a song and verse tell us that the Moon has soul). The word "dirt" is used colloquially, especially in the United States, for any combination of earth and soil, but it has no technical meaning. Some would restrict use of the word "dirt" to material that has been artificially moved, for instance, on a tractor tire, the sole of a boot, or below fingernails. And because in America "dirt" also refers to gossipy information, if you want vineyard puns then the word is an endless source.

Some other everyday colloquial words have no technical meaning. For example, we often talk of an easily visible *fragment* of Earth's rock material as being a "rock" or a **stone**. They are pieces or chunks, and they may consist of a single mineral, but more than likely they are bits of geologic rock, of any kind. To some, a stone may be the smaller of the two; to others, a rock is more jagged. All this imprecision leads geologists to avoid using the words in this way, apart from certain specific applications such as road stone, building stone, or precious stone. For vineyard soils, however, describing them as rocky or stony immediately conveys a useful image.

In general, crops grow in the upper part of soil, the **topsoil**—moist, humus-rich, and hence fertile (Chapters 9 and 10). Most farmers do their best to get rid of any stones, and what's below the topsoil doesn't matter much to them. Actually, it's usually a compact sediment referred to as **subsoil**, although since it can't really support plant growth, it isn't technically a soil. But these distinctions are tricky with regard to vineyards. Some vineyard soils are legendary in their stoniness—prompting the question, "how can vines grow in *that*?" But there will be humus below the surface, out of sight where the roots are working, and some vines have roots probing deep into the subsoil and even into bedrock fissures, seeking water. Consequently, in viticulture the practical distinction between bedrock and the overlying soils is unusually blurred. So, when we return to the question "what is the ground of a vineyard made of?" we can say the following. Depending on how we look at it, it's made of soil with its moisture and humus, and it's made of rocks, stones, minerals, and the chemical elements and ions that underpin all of them. They all play a part.

Functional Yet Beautiful: Geologic Maps

By far the best way of showing the nature of the ground in a vineyard region and how it varies from place to place is with a map. Growers these days utilize highly detailed plots of specific soil properties, but generally wine commentators talk about vineyards on granite, limestone, slate, and the like—in other words the bedrock. These properties are effectively shown by geologic maps. These maps are extremely efficient documents that collate and communicate complex data in a single picture; yet, at the same time, they are hugely attractive. That artwork hanging on the wall of a winery's visitor area might actually be a geologic map!

Essentially, a geologic map is a plot of what geology is where at the Earth's surface. Normally, any agricultural soil, vegetation, concrete, and human-made debris that covers the land is ignored. It's as though we are looking through such covering materials and seeing the bedrock laid bare. Geologic maps appear in several chapters of this book.

Soil maps may seem a more relevant source of information, and many countries have maps of agricultural soil. Figure 1.4 presents some sources of further information, but these maps have never really caught on in the wine world. Part of the

Country	Portal
Viewing and purchasing geologic maps	
Australia	http://www.ga.gov.au/about/projects/resources/continental-geology
France	http://infoterre.brgm.fr/
Germany	http://www.bgr.bund.de/EN/Themen/Geodatenmanagement/Geoviewer/geoviewer_node_en.html
Italy	http://sgi.isprambiente.it/geoportal/catalog/sgilink/sgilink.page
New Zealand	http://maps.gns.cri.nz/
Spain	http://info.igme.es/cartografiadigital/geological/default.aspx?language=es
Switzerland	https://map.geo.admin.ch/
U.K.	http://mapapps.bgs.ac.uk/geologyofbritain/home.html
U.S.A.	http://ngmdb.usgs.gov/maps/mapview/
Europe	http://www.bgr.de/app/igme5000/igme_frames.php
World	http://www.onegeology.org/portal/
Soil classification systems	
International Union of Soil Sciences	http://www.iuss.org/
FAO-UNESCO soil system	http://www.fao.org/soils-portal/soil-survey/soil-maps-and-databases/faounesco-soil-map-of-the-world/en/
Soil maps	
England and Wales	http://www.landis.org.uk/
France	http://www.gissol.fr/
Australia	www.csiro.au/en/Research/AF/Areas/Sustainable-farming/Decision-support-tools/SoilMapp
New Zealand	https://smap.landcareresearch.co.nz (needs login)
U.S.A.	http://websoilsurvey.sc.egov.usda.gov/App/HomePage.htm
Archived maps of numerous countries of the world	
	http://esdac.jrc.ec.europa.eu/resource-type/maps

Figure 1.4 Some examples of portals that give access to viewing and purchasing geologic and soil maps.

problem is the unusual nature of some soils in which vines are able to grow, as mentioned earlier. Also, there is no international consensus on how to classify soils in a practical way. For example, names based on places in the United States, such as Bale, Boomer, and Jory soils, are difficult to apply elsewhere; Australian groupings of soil properties are generalized and hard to relate to individual sites; and academic groupings such as andisols (or andosols in a different classification), cambisols, and leptosols are seldom seen in popular writings. So I'm talking here about the basic, widely available, general geologic maps (which outside the United States are generally called "geological" maps) that are most commonly seen in the general wine context.

Ever since 1815, when William Smith singlehandedly produced his momentous map of England and Wales—which Simon Winchester dubbed "the map that changed the world" and became the inspiration of an anthology of no fewer than 42 poems about the map—numerous individuals, learned societies, and commercial organizations have produced geologic maps according to need. However, the official geological survey of a region or country is normally the first port of call when one is acquiring a map of an area of interest (Figure 1.4). Most official surveys now have a strong web presence, including catalogues of maps and facilities for online ordering, and a surge has taken place to digitize existing maps and make them directly available online.

One advantage of digital maps is the relative ease of keeping them up to date: some paper maps that are routinely available for purchase are over half a century old! All the same, the early maps, water-colored by hand, are some of the most sought-after artifacts in geology. An original of William Smith's great map, for example, was recently on the market at a price equivalent to several cases of Chateau Petrus (a single bottle of which, even for the most recent vintage, sells for well over $5000). Photographic reproductions of the map are readily available, but there is no chance of actually reprinting Smith's map from the original copper plates. When put up for sale in 1877 "for trifling cost," nobody wanted them. So they were melted down.

Probably the most striking aspect of a typical geologic map is its numerous swaths of color (Figure 1.5; see Plate 2), or black and white **ornament** on an uncolored map. These indicate the distribution of the **map units** or **map formations** into which the earth materials have been divided for the purpose of the map. Loose geologic materials overlying bedrock are often referred to as **surficial** or **superficial** deposits and may be shown on larger-scale geologic maps if they have substantial development, say a few meters or more in thickness. This will be the case in vineyard regions located on thick sands and gravels, loess, and the like—such as the Médoc, the Napa benches, or the Langenlois in Austria.

All outdoor enthusiasts will be familiar with topographic contours—lines joining points of the same elevation (height above sea level) and hence portraying the ups and downs of the land surface. Some geologic maps have analogous contour lines that show variations of things *below* the ground surface. Geologists call them

Figure 1.5 Examples of geologic maps. (a) The Côte d'Or, France, between Chambolle-Musigny northward to Gevrey-Chambertin. The pale cream colors are Tertiary and Quaternary deposits, such as those of scree (E) and loess (S). The brown, gray, and blue colors are Jurassic bedrock, including Comblanchian limestone (j2c), an iron-bearing oolitic limestone (j4), and Oxfordian-age marls and limestones (j6). Extract of geologic map no. 499 (Gevrey-Chambertin) at 1:50,000 (© BRGM, Authorization R16/21). (b) The Barossa Valley region, South Australia. Vines are grown most intensely on the superficial deposits (Qpp, colored cream on the map) that form the valley floor (central on the map), which is flanked on both sides by a variety of rocks of Tertiary (T), Cambrian (ϵ), and Precambrian (P) age; black lines are faults, such as that defining the northeast of the valley, around Stockwell. Reproduced from the 1:250,000 Adelaide geologic map GM-SI54/9, by permission of the Geological Survey of South Australia, South Australian Department of State Development.

structure contours, and they are simply contour lines portraying some surface that happens to be beneath the ground, such as the top of the water table or the depth to bedrock (Figures 1.6 and 1.7).

When seeking the best sites for viticulture in an area, the conventional approach is to consult maps to get at least a reconnaissance idea of the area's geology and soils (although someone once said that in Europe you look for a monastery and then the spots where the winter frosts melt first). A geologic map is the quickest way to find out about the geology of an area, but it will probably be of limited use for

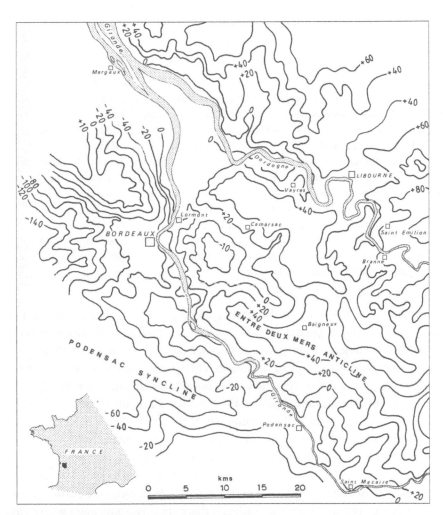

Figure 1.6 Example of a structure contour map, Entre-Deux-Mers, France. Here the contours show the altitude of the top of the Calcaire à Astéries formation, the "starfish limestone" much mentioned in viticultural writings on this region. From Alex Maltman, *Geological Maps: An Introduction* (New York: Springer, 1990), p. 19. http://link.springer. com/book/10.1007%2F978-1-4684-6662-1, with permission of Springer.

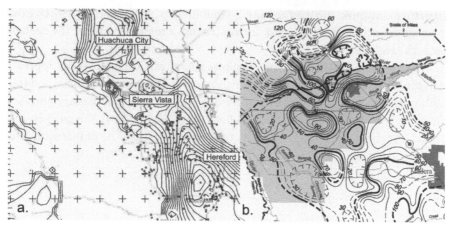

Figure 1.7 More examples of structure contour maps. (a) Contours of the depth to bedrock at Sierra Vista, a developing vineyard area near San Pedro, in southeast Arizona. In the strip running southeastward from Huachuca City through Sierra Vista to just west of Hereford, the densely packed contours show that bedrock is buried by a kilometer and more beneath the ground surface. In contrast, the strip to the west of Huachuca–Sierra Vista (where the wineries are concentrated) lacks contours, indicating that the bedrock is only tens of meters deep and thus may conceivably be reached by vine roots seeking water. The tiny circles are the locations of the drill-hole data on which the contours were based. From USGS Open File Report 00-138. (b) Contours showing the height of the water table in California's Chowchilla groundwater basin, in the Madera AVA. The area produces around 10% of the state's wine grapes, thanks to irrigation. With these contours and knowledge of the land elevation, a grower can estimate the well depth to tap groundwater for irrigation, and hence obtain a first approximation of the likely cost.

more detailed matters such as the actual siting of a vineyard. For one thing, published geologic maps are unlikely to be at a sufficiently large scale. A typical map at 1:50,000 or smaller is not going to provide much detailed information, and anything much larger is likely to be of an urban area. (Map scales are ratios, where one unit of distance, be it inches or centimeters, on the actual map represents the specified number of the same distance units on the ground. So 1:50,000 means that 1 centimeter on the map depicts 50,000 centimeters; that's 500 meters or half a kilometer.) Unless the vineyard covers an exceptionally wide area, it's likely to be represented by only one or two map divisions; variations will simply not be indicated. A particular vineyard in La Clape, in the Languedoc, for example, is shown on the standard geologic map to be entirely underlain by limestone. However, vine performance is extremely variable across the vineyard, and there are even some parts where vines have trouble surviving. At a much larger scale, however, it turns out that there are crucial variations in the nature of the limestone bedrock.

Then there is the problem that because geologic maps show the bedrock (and soil maps tend to emphasize the soil at the very surface), the bulk of the material that is actually around the vine roots is not represented! The geologic map of Bonnezeaux in the Loire Valley shows that the bedrock in the lower Layon area is schist; it does not show that the overlying soil variously comprises fragments of volcanic rock and of sandstone and shale derived from the higher upslope. Nor does it show that the soil varies rapidly in thickness from less than 0.2 meters to more than 1.2 meters. The soil map of Coonawarra, Australia, shows the famous terra rossa soils (typically, only half a meter thick), but the key to their suitability for vines is the drainage and water storage provided by the underlying limestone bedrock.

To the rescue, however, have come new technologies that allow the fine-scale surveying of properties directly relevant to the grower, techniques referred to as **precision viticulture**, or simply **PV**. This is not geologic mapping in any conventional sense, although it can involve the same remote sensing methods now utilized in surveying rocks. William Smith traversed the country alone on horseback, but today much information is gained by sensors that are not even on the ground, but overhead in a drone or aircraft and, most commonly, mounted on a satellite. Data can also be collected from within the vineyard itself, from geographic information system (GIS)-pinpointed measuring devices, sentinel vines equipped with sensors, and analysis at the time of harvest. The power available to enable the grower to micromanage the vineyard is already staggering and is growing rapidly. At Monthélie in Burgundy, for example, soil-surface variations have been mapped using aerial imagery at a scale down to five centimeters. Yes, 5 centimeters!

Therefore, for such hands-on purposes, the conventional geologic map now has to be just a starting point, an initial framework on which to hang the plethora of practical data that have become achievable. Of course, geologic maps remain of more general interest to wine lovers and provide attractive illustrations in wine literature. Can there be any scientific document that is more handsome?

The Big Backdrop: Plate Tectonics

No modern account involving geology can ignore the two revolutions in thinking that not long ago suddenly made it possible to see how minerals, rocks, and the processes that form them all fit together in a grand overall Earth scheme. First, geologists increasingly realized just how internally mobile the entire planet is. That fine writer on geology, John McPhee, summed it up in his 2000 book "*Annals of a Former World*" while talking about Mount Everest: "When the climbers in 1953 planted their flags on the highest mountain, they set them in snow over the skeletons of creatures that had lived in a warm clear ocean . . . possibly as much as 20,000 feet below the sea floor. This one fact is a treatise in itself on the movements of . . . the Earth." We are still discovering just how restlessly mobile the interior of our planet is.

Moreover, geologists stopped thinking of a series of roughly concentric and independent shells like the solid Earth, the oceans, the biosphere, and the atmosphere, but recognized an interacting, dynamic "Earth system." As one example, a magnetic field is generated deep within the Earth, in the core, and this affects rocks, the oceans, and the life patterns of some creatures. It extends to the outer reaches of the atmosphere, for instance, making the wondrous lights at the North and South poles. In a way, the interaction in a vineyard of bedrock, soil, organisms, and climate is a microcosm of this dynamic Earth system.

The second revolution involved the dawning of "plate tectonics." This new theory isn't directly relevant to growing vines, but it's mentioned in many a report on wine regions, so here is a brief outline of the principles. Geologists used to see the Earth's crust as pretty much stationary and fixed, apart from some up-and-down movements to account for mountains, and, of course, the oceans had always been there. Then about half a century ago, it emerged that the outer part of the solid planet is segmented into a number of slabs called **tectonic plates,** or simply **plates**, with each plate sliding differently in continuous, slow, *sideways* movements. Geologists suddenly realized that this concept explained a great deal about the features of the physical world around us, and it was soon labeled **plate tectonics**. (The word "tectonic," also used by architects, implies a connection between Earth's structure and building: indeed, the Greek word for mason is *téktonas*.)

The plates move ultimately because the Earth has internal heat. Such a "power of heat" within the Earth was first recognized by none other than James Hutton, but back then his ideas were controversial. He speculated that the Earth was a giant heat engine, but he had no idea where the heat came from. In the modern scientific view, it has two sources: a vestige of the temperatures resulting from cosmic matter accreting to create the Earth in the first place and, much more importantly, the heat continually thrown out by the ongoing disintegration of the planet's natural radioactive elements. The plates move rather like scum on the surface of a saucepan of soup being heated on a gas ring. The liquid immediately over the gas flame heats quickest and starts to rise against the adjacent, less hot material. Before long, it's moving sideways on little cells of moving soup.

Figure 1.8a shows the plates as they are at the present time and the direction of their relative movements. Of particular interest are the zones where the plates meet—**plate boundaries**—because here things are particularly dynamic. There are three main situations (Figure 1.8b). In the first situation, two plates come together, one driving below the other, and so we have a **convergent** plate boundary. Fractures can splay off upward from the zone that separates the two plates, slicing off parts of the solid ocean floor and its sediments, plastering them onto the edge of the overlying plate. Just such a process took place in the geologic past at what is now the central California coast. This explains why slivers of limestone, much prized by some growers in the region, occur in places within a complex mass of basalt and other rocks that represent the bedrock of an ancient ocean floor.

Plates move just a few centimeters a year or so, a rate that is almost imperceptible to us. So converging plates are hardly "crashing together" or "colliding," as it

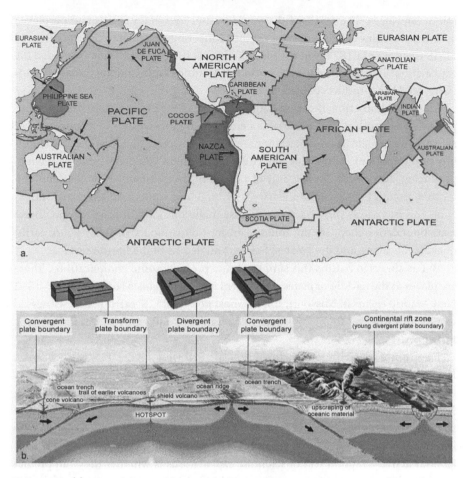

Figure 1.8 (a) Map of the Earth's tectonic plates as they are today. Arrows indicate the direction of movement. (b) Cross section showing the main kinds of plate boundaries. Adapted from artwork by Jose F. Vigil, in *This Dynamic Planet* (U.S. Geological Survey, Smithsonian Institution, and U.S. Naval Research Laboratory, 2013).

is sometimes put; rather, they are slowly grinding past each other, at a speed comparable to our fingernails growing. Active plate convergence today forms the backdrop to a number of wine-producing regions: those in South America stretching from Lima in Peru down to Bio Bio in Chile are located where the Nazca and South American plates are converging. This explains why the region is prone to earthquakes and bordered to the east by a chain of over 200 volcanoes.

A second situation involves plates moving apart, forming a **divergent** plate margin. As the plates separate, molten rock rises from below to continuously fill the space, thus balancing the material consumed at convergent boundaries and allowing the planet to maintain its constant size. This movement happens particularly in the central parts of oceans, where the accumulations of the rising rock form the Earth's midocean ridges. The origin of these underwater mountain chains had perplexed scientists ever since their discovery on the pioneering oceanographic expedition of HMS *Challenger* in 1872. Almost a hundred years later plate tectonics provided the explanation. The basalts of the Azores in the Atlantic Ocean and the Canary Islands (Chapter 4) are products of plate divergence.

In the third situation, there are boundaries where the plates are horizontally sliding past one another, and material is neither being formed nor consumed. These are **transform** plate boundaries. They are less widespread than the other two kinds but very important in the wine world. As discussed in Chapter 7, the main wine-producing districts of Turkey and of northern South Island, New Zealand, are located around transform boundaries, as is much of California: the San Andreas Fault and its various splay fractures are essentially this kind of plate margin. These plate boundaries, then, are the zones of greatest dynamic Earth activity—of earthquakes, volcanoes, and the greatest heights and depths of the Earth's surface.

But as always in nature, this simple picture presents some complications. There are places in the middle of plates that suffer large earthquakes (e.g., the 1811–1812 New Madrid events in Missouri, still the most powerful U.S. earthquakes witnessed east of the Rocky Mountains) and volcanism. We call these volcanic locations **hot spots**.

Far from a plate boundary, the furthest northwest of the Hawaiian Islands, Kaua'i, is the oldest volcano in the chain, having erupted over 5 million years ago. O'ahu, the island with Hawaii's capital, Honolulu, is about 2 million years old; Maui, further southeast, is about 1 million years old, and the Big Island, furthest southeast, is erupting presently. We therefore infer that the Pacific tectonic plate is moving northwestward over a fixed hot spot below it, thus producing a trail of igneous activity. A good test of scientific explanations is whether they can predict events, so here we might foresee a future volcanic island to the southeast of the Big Island. And guess what! To the southeast, nearly a kilometer beneath the waves, is a currently erupting volcano, Lō'ihi, looking as though it will form the next Hawaiian island—in 50,000 or so years' time.

Of more viticultural relevance than Hawaii (though it does have three wineries) is the hot spot below the northwestern United States inferred to account for the ages of the region's abundant basalts. In this case, it is the North American tectonic plate that is moving southwestward over a fixed hot spot. Outpourings began about 16 million years ago in what is now southeastern Oregon. About 12 million years ago lava was erupting to the northeast in Idaho (south of Boise, in part of the Snake River Valley AVA) and by 8 million years ago in the Eastern Snake River Valley just north of Pocatello. Today, the hot spot is thought to lie underneath the high volcanic plateau that is Yellowstone, prompting worries about future eruptions there.

A hot spot trail has recently been located in eastern Australia, running from the central Queensland coast south through Riverina into the Goulbourn Valley wine region, Victoria, and beyond. The heat focus currently seems to be just north of Tasmania. No volcanoes have erupted on mainland Australia since Europeans arrived there in the seventeenth century, though aboriginal people are likely to have seen eruptions in Victoria several thousand years earlier than that. This trail of hot spot volcanoes is far from a plate boundary but stretches over 2000 kilometers. This makes it the longest known volcanic chain on Earth. The recency of its discovery provides a good illustration of how we are still realizing the ongoing dynamic nature of our planet.

Further Reading

A simple guide to general geology, written for the beginner is:

Park, Graham. *Introducing Geology: A Guide to the World of Rocks*. Edinburgh, Scotland: Dunedin Academic Press.

Two standard university-level introductions to geology (but not at all about wine), both of them authoritative and beautifully illustrated, are:

Grotzinger, John, and Jordan, Thomas. *Understanding Earth* (7th ed.). San Francisco: W. H. Freeman, 2014.

Marshak, Steve. *Earth: Portrait of a Planet* (5th ed.). New York: W. W. Norton, 2015.

The concept of terroir (discussed in the present book in chapter 10) is usually taken to encompass the geology of a vineyard together with its various climatic aspects, and, to some writers, further things such as microbiology and human traditions. However, although the following two books are entitled "Terroir" they only cover geology.

Wilson, James. *Terroir: The Role of Geology, Climate, and Culture in the Making of French Wines*. Oakland: University of California Press, 1999, and

Fanet, Jacques. *Great Wine Terroirs*. Oakland: University of California Press, 2004.

The book by Wilson is authoritative and readable, and Fanet's is beautifully illustrated and does add some parts of the world outside France. Both, however, are descriptive accounts of the geology and don't address the issue of how it affects vines and wines.

Robinson, Jancis, and Harding, Julia (eds.). *The Oxford Companion to Wine* (4th ed.). Oxford: Oxford University Press, 2015. This book has numerous separate entries concerning geology, minerals, rocks and soils in vineyards.

2

How Minerals Work

We might expect the ground of vineyards to consist of bewildering permutations of elements, but because its composition is dominated by just eight of them and there are chemical restrictions on how they can combine, the number of common minerals is not huge. Even so, their names are not particularly well known, even those of the very minerals that make the ground we live on and the soils that vines grow in. Mineral names that might spring to mind are more likely to be those used in jewelry or that are commercially mined. Such **gemstones** and **ore minerals** are not widespread, but geological processes have concentrated them in certain parts of the Earth, and if we can locate these accumulations, it may be worthwhile to extract them for profit.

The rocks and soils that mainly concern us here are composed of silicate minerals, and Chapter 3 is devoted to these workhorses. They are sometimes called the "rock-forming minerals," though there is an outstanding exception to this term: calcium carbonate, which makes the calcareous rocks. So in this chapter's survey of the kinds of nonsilicate minerals we may come across in vineyards, we will pay particular attention to the carbonates. But first, let's examine some fundamental concepts concerning the nature of minerals.

Minerals as Crystals

As we saw in Chapter 1, minerals are made of ions bonded together through giving or sharing electrons. But to achieve this linkage, the ions cannot combine in some higgledy-piggledy fashion; rather, they have to organize themselves in a particular, symmetrical physical arrangement. It's a bit like the sight of soldiers on formal parade. We call the three-dimensional framework of ions a lattice, and it's this regular pattern that makes the material **crystalline**; it is a **crystal**. In other words, the pieces of mineral in a vineyard are crystalline. We may think of crystals as having the attractive, light-catching facets seen in gem shops and museums. Although this is a manifestation of the crystalline structure of the constituent ions, it is not what defines them as crystals. Consequently, minerals lying in a vineyard may be dull, shapeless chunks, but *they are still crystals.* The accompanying box explains what determines whether or not a piece of mineral has a good crystal shape.

So, saying that the soils in a vineyard are "crystalline" has no real geological meaning. All vineyard soils are composed of crystals of one sort or another, not some more than others. Just because some of the soil particles happen to have smooth external faces and catch the sunlight, thus making the soil sparkle, this doesn't make it any more crystalline. Also, normal liquids, including wine, have no crystal structure. So although wine is sometimes described as having a "crystalline" aspect, this has to be purely metaphorical, unrelated to the physical properties of the wine or of geological materials in the vineyard.

The Shape of Crystals

The example of a crystal structure shown in Figure 2.1 is the mineral quartz. The silicon and oxygen ions make a distinct pattern, but whether this is manifested as smooth crystal faces on a piece of the mineral depends on the circumstances in which it formed. Essentially, to develop nice external facets, a mineral has to

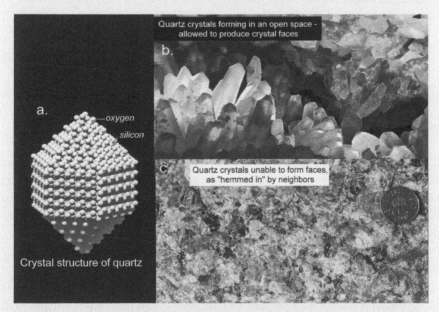

Figure 2.1 Crystal structure versus crystal faces. Almost all minerals are crystalline, as in the systematic arrangement of silicon and oxygen ions seen in (a). Natural crystal faces replicate this crystalline structure, but the extent to which they are allowed to develop depends on whether the crystal is growing into spaces, as in (b), or is constrained by adjacent crystals, as in (c).

"grow" during its formation into a space and not be constrained by adjacent miner-
als. Take, for example, a cooling volcanic lava. Its constituent ions will be moving
around, uncombined, in this runny state. But as the lava cools and begins to solidify,
appropriate ions meet together and form a clump—a nucleus—onto which fur-
ther suitable ions join. Each ion locks on to a site in the lattice suitable for its size
and electrical charge, creating the regular arrangement characteristic of a crystal.
Geologists say that in this way the mineral "grows," though, of course, there is noth-
ing biological involved. The outer boundaries of this growing mineral will tend to
reproduce regular surfaces in the crystal lattice: it is at that stage that it will have
crystal faces.

With continuing cooling, however, inevitably the growing minerals will start
to interfere with each other, and this starts to restrict the development of regu-
lar external faces. Eventually, the lava wholly solidifies: the mass of minerals
becomes interlocked to give an igneous rock. Each of the minerals in the aggregate
is a crystal, but hardly any will have been able to complete development with
unrestricted crystal faces. Each will have been too "hemmed in" by its neighbors.
But should there have been, say, a gas bubble in the lava, then a mineral growing
into it may have had the opportunity to form unrestricted external faces. (The
same thinking lies behind the perfect cubes of halite, common salt, crystallizing
on the shores of the Dead Sea, as featured in a number of on-line videos.) In other
words, the development of smooth external boundaries in a mineral requires spe-
cial circumstances.

A Span in Composition

As a direct result of their crystalline nature, some minerals have a considerable
range of chemical composition, within fixed limits. As a mineral is forming, if there
are two different elements available that happen to have similar sizes and electrical
charges, then either of them could fit into the crystal lattice, leading to a span of possible compositions.

For example, the mineral calcite is calcium carbonate *if it is pure*. However, the
magnesium ion has rather similar properties to calcium and is able to substitute for
it in the crystal lattice. So often in nature there is a certain amount of magnesium in
calcite, and if it is significant, geologists talk about "high magnesium calcite." If as
many as half the sites in the crystal are occupied by magnesium instead of calcite, the
mineral is sufficiently different for a new name to be justified: dolomite. If virtually
all the sites in the carbonate lattice are occupied by magnesium instead of calcium,
then the mineral is different again. It is called magnesite, and it completes the range
of possible calcium/magnesium ratios. Such spans of composition are common in
minerals, as we shall see in the next chapter with the silicates.

Identifying Minerals

If you're visiting or working in a vineyard and you see what may be fragments of a mineral, how do you go about identifying them; that is, how do you give them a specific name? Unfortunately, often the most obvious attributes of minerals aren't very helpful. The substitution of one element for another in the crystal lattice, as mentioned earlier, together with the possibility of trace impurities, mean that a particular mineral can have a variety of appearances; the chemistry that defines a mineral doesn't always correlate neatly with the outward appearance.

Color is a case in point. Although color readily helps us distinguish between, say, a pied wagtail and a yellow wagtail, or a red mullet and a gray mullet, a particular kind of mineral can have a whole range of colors. A few minerals, such as the mineral turquoise, are practically defined by their color, but most—and certainly those we are likely to encounter in vineyards—can vary greatly in color. Even very tiny amounts of impurity can have an enormous effect (e.g., see quartz in Chapter 3). Thinking back to our parading soldiers, we see that the display would have a very different overall look if one or two of them were in the wrong uniform!

In 1811, Friedrich Mohs was busy surveying the rocks in the Southern Styria wine region of Austria when it dawned on him that minerals had certain systematic *physical properties*, irrespective of color. He realized, for example, that the hardness of a mineral was consistent, and he arranged the minerals in his collection—some now on display in Vienna's Natural History Museum—in order of decreasing scratchability. Today, every geology student learns about the "Mohs scale of hardness," as such properties, arising from the regular internal crystalline structure of a given mineral, provide a great shorthand aid to identification. Books on minerals catalogue the most useful properties, but not always with explanations, so the accompanying box outlines two of the most useful ones for vineyard soils: mineral hardness and mineral cleavage.

Identifying Minerals by Hardness and Cleavage

The mineral gypsum can look superficially like quartz, but, as minerals go, it's rather soft. So a piece of a whitish, glassy mineral that you can dig your fingernail into is a good clue that it might well be gypsum. Quartz, in contrast, is noticeably hard—even a knife blade won't scratch it. (Together with its chemical stability, it's this quality that makes the mineral tough, robust, and virtually insoluble.)

Minerals can be arranged in order of increasing hardness—Friederich Mohs's scale—meaning that any substance higher in the list can scratch a lower one. But care is needed if you're using this property! You have to make a definite scratch in the softer mineral, preferably checked by being able to blow away the loosened material. A softer substance drawn across a harder one can leave a mark (from the softer material), but it's not a scratch. Testing quartz with a knife, for example, can leave a dull gray mark. Also, you have to be sure you are testing the mineral itself and not a layer of some weathered, decomposed material on its surface—a common situation for some minerals lying around a vineyard. And it has to be a single grain of mineral. If the sample is an aggregate of fine grains, then scratching it is more an assessment of how well the grains are bonded together, which is not the same thing as mineral hardness.

*Then there is **mineral cleavage** (Figures 2.2 and 2.3). A crystal lattice may be such that the bonds between the ions are less strong in particular orientations, giving the mineral a tendency to break, or "cleave," in these directions—it will have "planes of cleavage." An extreme example is the mineral mica (Chapter 3). Because the ions in its crystal structure are arranged in sheets, with relatively weak bonds between them, mica shows a marked tendency to split into these leaves. In a vineyard soil, the mica will appear as thin plates. If they are big enough, it will be easy to peel them apart, utilizing its prominent cleavage.*

There can be more than one direction of weak bonding within a crystal lattice; the mineral feldspar, for example, has two cleavages (though they are not prominent). The cleavage directions can be more or less at right angles to each other, as with feldspar, or they can be oblique. The mineral halite (common salt), for example, has three cleavages at right angles. Hence, it breaks into cubes (in line with how

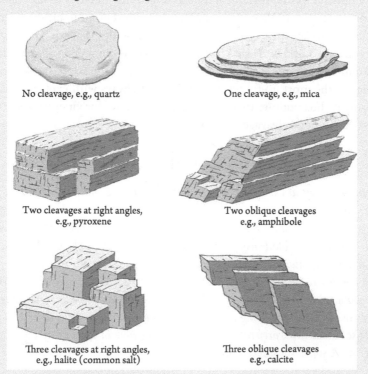

No cleavage, e.g., quartz

One cleavage, e.g., mica

Two cleavages at right angles,
e.g., pyroxene

Two oblique cleavages
e.g., amphibole

Three cleavages at right angles,
e.g., halite (common salt)

Three oblique cleavages
e.g., calcite

Figure 2.2 Sketches of mineral cleavage.

Figure 2.3 Photographs of mineral cleavage. (a) The single cleavage of mica.
(b) Two cleavages in pyroxene (augite). Along the length of the mineral, the two
cleavages make indistinguishable traces, but across the end they are seen as two
planes at right angles. (c) The three cleavages of halite, common salt, at right angles,
make the mineral break into cubes (compare with Figure 1.2). (d) The two cleavages
of feldspar, at right angles, which tend to make the mineral break into box-shaped
fragments.

*it grows). Calcite also has three directions of cleavage, but these are oblique, at 120°
to each other. It isn't usually necessary to actually break the mineral to assess any
cleavage because often there are fine traces looking like hairline cracks and steps
running across a mineral grain, representing incipient surfaces of fracture.*

*In contrast to all this, the crystal structure of quartz has no particular weak
direction: quartz has no cleavage. There are no particular directions along which a
grain of quartz will tend to break (see Figure 3.4). However, the random planes of
fracture will be distinctively curving, much like the way window glass breaks, properly called* **conchoidal fractures***. There is a belief, very fashionable in some quarters, that the performance of vines is improved by spraying on them finely ground-up
quartz. The grinding is supposed to multiply the number of reflective crystal facets
because (in a way unknown to science) this amplifies something called "cosmic
light.". This notion comes from Rudolf Steiner, a messianic figure to some winemakers. Steiner lectured on things ranging from the Apocalypse to Zarathustra, and in*

a 1924 talk on the Ego and Meditation pronounced: "Take a quartz crystal. If you take a hammer and break it up, the single pieces retain the tendency to be six-sided prisms, capped by six-sided pyramids." (He went on to assert that from the smallest pieces "something living and cosmic" would emerge). But quartz crystals don't do this; demonstrably, they break into shapeless grains. In fact, this is a further way in which quartz can be distinguished from that other common whitish mineral, calcite, with its three cleavage directions. Other materials that show conchoidal fracture are the mineral flint (Chapter 3) and the rock obsidian (Chapter 4).

Intricate but Vital: Cation Exchange

We turn now to a process that is vital, literally, for the growth of vines by making essential mineral nutrients available from the clay and humus in soil. In fact, the late soil scientist Nyle Brady once remarked that "next to photosynthesis and respiration, probably no process in nature is as vital to plant and animal life as the exchange of ions between soil particles and roots." (Of course, most animals depend at least indirectly on plants.) So it's perhaps surprising that ion exchange was discovered as recently as 1930.

Most of the mineral nutrients that a vine needs are cations (Chapter 9). Some nutrients may be gained directly through the tiny mycorrhizal fungi that live along with the vine's root hairs, but otherwise they come largely by humus and clay minerals making some of their ions available to the roots. The ability of a material to do this is expressed by its **cation exchange capacity** or **CEC**. It's a property sometimes alluded to in wine writings, but less often discussed is what it actually means, how it happens, and why the clay minerals do it dramatically more than others. I explain it in the accompanying box. Although wine writers like to remark on the importance of clays for vine nutrition, and they may even mention CEC, we shall see in Chapter 9 that in practice it's the humus that's more important, indeed essential. Its contribution to vine nutrition is absolutely vital, but wine literature gives little consideration to rotted organisms!

Explaining Cation Exchange

Rarely in nature do the constituent ions of minerals share all their electrons such that there is overall electrical neutrality. Minerals are usually left with some superfluous charge on their outermost surfaces, and this is particularly the case in the clay minerals. They have complex internal sheet arrangements, as we will discuss in Chapter 3, and among other things it's common for aluminum, Al^{3+}, to take the place of some of the silicon Si^{4+}. This gives a reduction in the positive charge, so that fewer negative charges are needed to balance it. The resulting superfluous electrons just drift around the surface of the mineral. The whole effect is vastly enhanced in *clay minerals because of the enormous cumulative surface area that results from their characteristic submicroscopic grain size (Figure 2.4).*

*So what do these clouds of electrons, these negative charges, do? Let's imagine that a nutrient such as potassium is in the soil pore water, perhaps originally derived from the breakdown of some mineral but now dissolved as K^+ cations. The tendency will be for it to be steadily leached away with the moving pore water and for it to be lost to the soil: what is needed is a way of storing the potassium so that it is available as and when the vine needs it. This is where the clay minerals come in. The negative charges on their surfaces are able to attract and capture positively charged nutrients such as potassium. However, the ions are only loosely held: we say that they are **adsorbed** onto the surface rather than being **absorbed** to become part of the crystal lattice. They are constantly interswapping, such that a dynamic film of different ions surrounds the clay flake, competing for places on the mineral surface. A cation is easily displaced by another one if circumstances are right. It's something like iron filings being attracted to a magnet rather than being blown away, but being only temporarily held there, and not becoming part of the magnet.*

So the various cations in the soil water are, so to speak, squabbling to achieve linkage with the clay surface, and hence electrical neutrality. And there is a pecking order, which depends partly on the magnitude of their electrical charge: a more strongly charged ion will displace a lesser one and be held more strongly. Thus, for example, calcium with its two positive charges easily displaces sodium, Na^+. Also

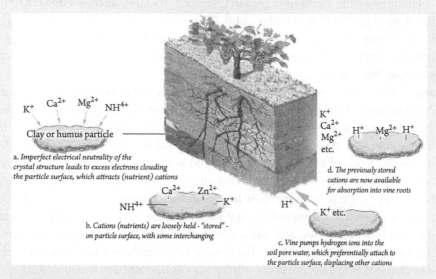

a. *Imperfect electrical neutrality of the crystal structure leads to excess electrons clouding the particle surface, which attracts (nutrient) cations*

b. *Cations (nutrients) are loosely held - "stored" - on particle surface, with some interchanging*

c. *Vine pumps hydrogen ions into the soil pore water, which preferentially attach to the particle surface, displacing other cations*

d. *The previously stored cations are now available for absorption into vine roots*

Figure 2.4 Cation exchange. Sketches to explain how, due to imperfections and substitutions in the mineral crystal structure, free electrons at the surface of a clay or humus particle attract and store cations (a and b). Hydrogen ions issued from the vine can displace the cations (c), making the cations available as potential nutrients (d).

important is the size of the ion: the smaller the radius of the ion, the more closely and strongly it is held. So in general, the order of decreasing tendency to link with the clay surface, taking into account both the charge and size effects, is Al^{3+} > Ca^{2+} > Mg^{2+} > K^+ = NH_4^+. (NH_4^+ is the ammonium cation, derived from ammonia, NH_3, and the vital source of nitrogen for the vine.)

But here's the cunning thing. The ion that makes the strongest link of all is hydrogen (H^+)—and it's almost as though the vine knows this! The vine's metabolism can prompt its roots to pump out hydrogen ions into the soil water, which then dislodges the other ions held on the clays, thus making them available to the vine roots. The vine has to use up energy to do this pumping, but it's advantageous on balance because the roots can now take up the freed cation nutrients to promote further growth.

Some Minerals of the Wine World

Native Elements

Having established those principles, we begin our survey of the minerals most likely to be encountered in vineyards, starting with the nonsilicates (we come to the silicates, the "rock-forming minerals," in the next chapter). Notwithstanding all I have just been saying about minerals being compounds, we begin with a few examples that have just the right complement of electrons for them to exist naturally with little or no tendency to combine with other elements. They are often referred to as the **native elements**.

I mentioned the inertness of *gold* in Chapter 1. It's rarely acknowledged that the discovery of gold in 1848 at Sutter's Mill, El Dorado County, California (amazingly, the actual tiny flake is preserved in the Smithsonian Museum) was a major trigger for beginning the state's wine industry. Even by the early 1850s, local wineries were supplying the gold miners; by the 1870s, El Dorado County was California's third largest wine-producing area. Interestingly, vines were first planted in New Zealand's Otago region in 1864 by a French prospector who was there primarily to seek gold.

Native *carbon* occurs in greatly differing physical natures. In some forms, the carbon atoms are disorganized—there is no crystal structure—and this includes the various kinds of coal. Being black, carbon materials such as coal dust, lignite, and soot have sometimes been sprinkled on pale-colored vineyard soils to darken them, supposedly to improve their heat-absorbing properties. Some sites in the Champagne region, for example, contain impure lignite that was long used for this purpose; in earlier times, coal waste was brought from the nearby mines around Lille to darken the vineyard soils. The noncrystalline carbon, often loosely referred to as charcoal, also finds use in the winery as a filtering and decolorizing agent.

Unspectacular though it may be, the most widespread occurrence of native carbon is in the tiny black specks scattered through some sedimentary rocks, the final remnants of some once living organisms such as marine plants. This is the origin of the dark color in black shales in vineyard soils (Chapter 5), such as those in the valleys of the Duruji River near Kvareli, Georgia, and of the Eel River, southern Humboldt County, California (in the heart of Redwood Country, visited by driving along the Avenue of the Giants).

Carbon also occurs in crystalline forms such as **graphite**, named from the word in classical Greek for drawing. Most commonly, despite its regular internal ordering of the carbon atoms, graphite looks like shapeless clumps and smudges. They can be tiny. It's the dispersed fine particles of graphite that give the dark color to black slates and schists (Chapter 6), the metamorphic equivalent of black shales. Graphite can also form with crystal faces, a flat, plate-like form expressing the internal arrangement of the carbon atoms. This is why the graphite-bearing rocks and soils of the Priorat region of northeast Spain are often described as "sparkling" in the sunshine. Some of the vineyard soils in South Styria, Austria, have unusually high concentrations of graphite; in fact, the mineral is actively mined there.

Wines from such areas are sometimes described as tasting of graphite; indeed, the word is common in tasting notes, mostly for red wines. However, it has to be a metaphor, as the mineral graphite has no flavor. In 2016 a series of articles in the wine magazine Decanter attempted to "decode" tasting terms and tackled the "graphite and cigar box" notes by explaining that the term *graphite* signifies "notes of pencil lead or a lead-like minerality." This mention of lead seems unhelpful, seeing as pencils have never involved the element lead. Pencil "lead" is made of graphite mixed with kaolinite or bentonite clay (Chapter 3), and, just like metallic lead, it is both tasteless and odorless. The explanation concluded by saying that "if you are unsure what graphite smells like, try sharpening an HB pencil." The pronounced odor of old-fashioned pencils is due to highly aromatic compounds known as sesquiterpenes, which are found in most kinds of cedarwood, the material traditionally used for making pencils (and, incidentally, cigar boxes). These compounds are well known in wines, so this would seem to be the source of the "graphite" pencil perception; it can't be the mineral itself.

Most natural **sulfur** is found in association with volcanoes, usually in its powdery yellow form, such as around Vulture, Vesuvius, and other volcanic vineyards of southern Italy. Sulfur was mined in Roman times around Tufo in Campania (discussed in greater detail in Chapter 4). Some of the mines around the Sabato River, an area noted for its Greco di Tufo wines, closed only recently, and chunks of yellow sulfur dot the vineyards. In general, however, the sulfur in vineyard soils has been added artificially, for example, to combat mold.

A number of other chemical elements are sometimes mentioned as being significant in vineyards—**manganese**, for example, in Moulin-a-Vent, Beaujolais, France, or iron in terra rossa—but these are not present in a native form: they are combined with other elements in compounds. Thus, manganese exists in the ground combined with oxygen as the mineral called **pyrolusite**. Such combinations of elements with oxygen are called *oxides*.

Oxide Minerals

As the fourth most abundant element in the Earth's crust, it is hardly surprising that iron is almost ubiquitous, in some shape or form, in vineyard rocks and soils. Doubly charged iron, Fe^{2+}, is called **ferrous** iron, and iron lacking three electrons is **ferric** iron, Fe^{3+}. So converting iron from ferrous to ferric involves the loss of electrons, and such a reaction is known as **oxidation**. The element that has lost electrons has been oxidized: ferric is the oxidized form of ferrous. **Reduction** is the opposite. Of course, we talk about oxidation and reduction in wine, and while we think here of the effects of oxygen, sulfur, certain enzymes, and the rest, the chemical processes still fundamentally involve the loss and gain of electrons.

Hematite, sometimes spelled haematite, has iron in its ferric form. The mineral varies in appearance, but it's often associated with the color red. All the different forms of hematite give a dull red powder; in fact, the very name of the mineral comes from the Greek word for blood. It's perhaps this association that has led some wine writers to link its presence in vineyard soils with an enhanced red color in wine. However, the color of red wine is due to organic phenols such as anthocyanins, and not to iron in red soils.

The color of the celebrated terra rossa (Figure 2.5; see Plate 3) is due principally to clays stained with hematite (and goethite; see below). Such soils develop

Figure 2.5 Terra rossa. The striking red color of the soil is due to iron oxide, principally hematite, and hydroxide minerals, which leaching has left as a residue. Near Vodnjan, south Istria, Croatia.

in limestone areas with a Mediterranean climate, such as in Istria (Croatia), La Mancha (Spain), and Coonawarra (Australia). The iron-rich residue is left through limestone having been dissolved away. It provides sufficient nutrients for vines, while the limestone remaining below offers good drainage, which, together with the climate, may explain why good-quality grapes can be produced from these sites. These soils may have an eye-catching color, but there is no apparent justification for the special mysticism that some wine writers bestow on terra rossa.

Magnetite involves both ferrous and ferric iron. Its outstanding property is its magnetism; it is the only common mineral that will deflect a compass needle or attract iron filings. The ancients knew this and called it *lodestone.* Apparently, some of the vineyard pebbles on the slopes of Mount Majura, just north of Canberra, Australia, which derived from nearby volcanic rocks, effectively act as magnets because of their magnetite concentrations. The lavas to the west of the Mornington Peninsula, Australia, have yielded soils with a significant magnetite content; one wine produced there is called "lodestone."

Weathering of hematite and magnetite leads to a mix of minerals with many permutations of OH⁻ ions and water. In a word, the resulting materials are what we ordinarily call rust. In geology, they are collectively referred to as **limonite**. The group includes the ferric oxy-hydroxide **goethite**, named after Goethe, the German polymath celebrated for his drama "Faust" but also the enthusiastic owner of more than 18,000 geologic specimens. The color of these limonite minerals can range from the bright lemon yellow its name suggests to a drab gray-brown and, if hematite is mixed in, to a warm red-brown. These hydrous iron minerals have been used as pigments since time immemorial, such as in prehistoric cave paintings, and they are often referred to loosely as **ocher.** It is this coloring power that signals their presence in vineyard rocks and soils. Classical painters ground up these substances and then heated them vigorously to drive off the water, essentially leaving hematite. They called it burnt sienna, a term familiar from children's paint boxes.

The *ferrous* components in these various iron minerals tend to be soluble but the *ferric* compounds are not. Hematite can therefore be left as a residue when other materials have been dissolved away, exactly as in terra rossa. Conversely, should waters carrying dissolved ferrous ions become oxidized to ferric, then precipitates may arise. This explains the familiar yellow-brown deposits that smear water outlets, say, around old iron pipes. It is also why soil particles can become coated in a thin film of "iron" and take on a yellow-brown, ochrous appearance. Such precipitation can be a practical problem, not only in winery plumbing but also in irrigation systems in the vineyard. Amounts of ferric ion as low as 0.2 ppm can clog a drip irrigation system.

Bauxite, the common form of aluminum oxide, occurs as a leaching residue in soils of hot, dry areas, such as in the Pemberton wine region of western Australia. Long in the geological past, bauxite formed in such conditions in parts of what is now Provence, France. And in 1821, it was realized that the material around the hilltop village of Coteaux de Baux de Provence, northeast of Arles, could yield aluminum for commercial purposes, mines were opened, and the village gave its name to the mineral. That extraction is now virtually ended (though one quarry is now used as a tourist attraction in *son et lumière* productions!), but some of the vines lower down the hill still grow among bauxite rubble from the mining operations. Today, elsewhere in the world, large-scale strip mining coupled with the huge electricity demands of refining bauxite raise questions about the environmental impacts of bauxite mining, and hence about the use of aluminum screw caps by wineries presenting themselves as organic or sustainable.

Sulfide Minerals

A number of minerals comprising a metal combined with sulfur occur in vineyards. Probably the most widespread such mineral is the iron sulfide **pyrite**, which appears in the names of a number of wines from around the world. This is the well-known brassy-looking metallic mineral sometimes called fool's gold. It's a humble mineral but still important commercially; there is an argument that it spawned civilization. Its iron content enabled it to easily make sparks when struck (see pyrophoricity, Chapter 12), a capability that led to the use of fire and all that entails for cooking food, forging implements, and so on and so forth. Most commonly, pyrite occurs as tiny shapeless patches dispersed through a rock, but when well formed it makes little cubes, sometimes intergrown with each other. If the cubes weather away from the host rock, they leave neat little square holes, seen, for example, in some of the slates of the Barossa and Clare valleys, Australia. Chunks of pyrite are sometimes seen in loose sediment, such as on the lower slopes of Gevrey-Chambertin, Burgundy, in the tills of the Niagara region, and the volcanic soils of Santorini. In its Spanish form, *pirita*, it accounts for wine names in the Almaroja district of Arribes, along the northern edge of the Spanish border with Portugal, and in several places in Chile, including the Elqui and Leyda valleys.

    Galena is lead sulfide, a heavy, lead-colored metallic-looking mineral that often has a cubic aspect. In what is now northern Illinois, Native Americans used it in rituals and in body paint; as a source of lead, it was the location of the first of the many mineral "rushes" in the United States. The little town that sprang up as a result, Galena, is now the home of several wineries in the Upper MississippiAVA.

    Priorat, Spain, may now be synonymous with quality wine production, but for many centuries, in fact ever since prehistoric times, galena mining was the predominant occupation. Vine growing was a relatively minor activity. When the vines failed through the devastating attacks of the insect phylloxera in the late 1800s, it was mining that kept the area alive, with around twenty new mines opening at that time. But things have now come full circle. The last galena mine closed in 1972, and Priorat wines have experienced a renaissance. Today, several of the flourishing vineyards west of Falset overlie disused underground mine workings, and "galena" only lives on in winery names and wine labels. Zinc sulfide, the mineral **sphalerite**, is usually found along with galena, and the Priorat area has produced some famous specimens. It seems, though, that this word has less desirability as a wine name than its lead-bearing brother!

All the sulfide minerals mentioned earlier are celebrated in the Colline Metallifere, the metal-bearing hills of Maremma, southern Tuscany. While there is concern in some parts of this region about contamination from these mineral deposits, winemakers trumpet the qualities they supposedly confer on their wines. One estate boasts that its vineyard is located directly on top of an old pyrite mine, with the vine roots growing down into the roof of the mine.

Sulfate Minerals

Sulfate minerals have a cation combined with a sulfate, SO_4^{2-}, anion. They include **gypsum**, calcium sulfate, well known as plaster for encasing broken bones. Deposits of gypsum occur beneath Paris, particularly around the Butte of Montmartre (with its famous downtown Paris vineyard), where it was intensively mined in the Middle Ages for coating house walls. It acted as a fireproofing agent, which explains why Paris suffered far less fire damage than London. Thus, gypsum became known as **Plaster of Paris**.

Gypsum occurs naturally in some vineyard soils, such as around Murcia, Valencia, and Ribera del Duero in Spain. It has long been important around the town of Weinsberg, in the north of Württemberg, Germany, along with, as the town's name indicates, vines. The mineral was extracted from around a dozen quarries throughout the 19th century. One pit on the prominent hill of Burgberg has now been filled in and converted to a vineyard and is being used for experimentation by the State Education and Research Institute for Viticulture. Württemberg wines produced from gypsum-bearing soils are becoming fashionable, and the German word for gypsum, *gip,* is starting to appear on wine labels.

Plaster of Paris may be famous, but London, too, has contributed to the world of sulfates. In the seventeenth century, the waters of the spa town of Epsom, a few kilometers to the west of the city, became celebrated for their laxative properties. At the height of the vogue, people were apparently drinking up to 16 pints a day of the fashionable waters, with "various funny results." The secret ingredient of the waters? Magnesium sulphate, which still bears the time-honored name of **Epsom salts**. The compound is still used as a purgative—apparently the likes of Gwyneth Paltrow and Victoria Beckham swear by it—but also in viticulture. It is sometimes added to soils to increase their magnesium content and to wine must as a yeast nutrient.

Some wine writings confuse these various sulfur-bearing anions. So to be clear, *sulfates* contain four oxygen ions. They should not to be confused with either *sulfides,* which have no oxygen at all, or sulfites, SO_3^{2-}, which have three oxygens. All wine lovers know of *sulfites* and their preservative properties. They're almost routinely added to musts or wines, either as potassium or sodium metabisulfite (marketed to winemakers as campden tablets), as they readily dissolve and release the antiseptic gas sulfur dioxide, with its well-known pungent smell. It has been said that sulfites are the most important additive in the winemaker's medicine chest, but they are all manufactured, including those used in organic wine. Sulfites are not known in nature as minerals.

Carbonate Minerals

Carbonate minerals are defined by having a carbonate, CO_3^{2-}, anion. We usually think of the balancing cation as a single element, although, as explained earlier, other cations are likely to be present to some extent. With **siderite**, the defining cation is ferrous iron. It is found in iron-rich limestones such as those in the *causses* of the Cahors region, France, and they can be concentrated in siderolithic crusts (Chapter 9). **Magnesite**, a component of some limestone-derived soils such as those near Mittelwihr, Alsace, has magnesium balancing the carbonate anion— hence it consists of magnesium carbonate.

By far the most important carbonate mineral for vineyards, and indeed the most important nonsilicate mineral, is **calcite**, calcium carbonate. Calcareous materials are those significantly composed of calcite, and this applies to a host of vineyard rocks and soils, such as marble, travertine, tufa, marl, and all the various limestones (Chapter 5). Vineyards sited on calcareous rocks and the soils derived from them have acquired a particular cachet in the world of wine, and so the term appears frequently in the wine literature. Relative to the silicate rock-forming minerals, calcite shows a slight solubility in water. This is why limestone landscapes are so distinctive (Chapter 8) and why the majority of natural wine-storage caves are in limestone rather than other kinds of rock.

Calcite occurs naturally in a wide variety of forms. It can be clear with crystal faces, making little fragments glinting in vineyard soils. The three cleavage directions of calcite can also give smooth, reflective faces to broken fragments (Figure 2.2). Most commonly, a lump of calcite lacks much shape and is milky white, but small amounts of impurities can impart all kinds of shades and colors. It can look very similar in a rock or a soil to that other extremely common mineral, quartz. However, unlike quartz, it can be scratched with a knife blade, and it may be possible to discern the pattern of hairline cracks indicating the three oblique cleavage directions of calcite. Also, calcite will fizz with a dilute acid such as vinegar, as the gas carbon dioxide is released from the carbonate anion. If you know that a vineyard is sited on limestone, then the chances are that any whitish mineral fragments lying around will be calcite and not quartz.

Dolomite is a mineral closely related to calcite, but with half of the cations being magnesium rather than calcium. In many ways, dolomite looks just like calcite, but it fizzes with acid less vigorously. Dolomite is also the name of a rock, one composed almost entirely of the eponymous mineral. Hence, dolomite fragments of both minerals and rocks are found in the vineyard soils of Trentino and Alto Adige, in northeast Italy, the debris of the nearby Dolomites.

Historically, those spectacularly jagged mountains were vaguely called "pale mountains" by the locals, but an aristocratic French geologist doing fieldwork in the region realized that their carbonate composition was unusual. He reported his observations in 1791, and a year later the mineral and the rock—and before long, the mountains—were named after the nobleman: *dolomie* at first, in French, and within a few years, dolomite in English. His name, suitably grand, was Dieudonné Sylvain Guy Tancrède de Gratet de Dolomieu.

Further Reading

There are numerous popular guides to minerals and their identification, such as *Rocks and Minerals*, by Chris and Helen Pellant, New York: Bloomsbury, 2015, a beautifully illustrated introduction. Beware, though, as the foregoing chapter explains, of identifying such superficially variable things as minerals simply by seeking a match with a picture.

The mineral halite (common salt) crystallizing from the Dead Sea is shown in a video (together with a video explaining how salt dissolves as ions) at http://www.fromthegrapevine.com/nature/how-does-dead-sea-salt-form-giant-cubes.

3

The Minerals that Make Rocks and Soils

Seeing the Light—While on Vacation!

This chapter is about the minerals based on silicon and oxygen, the silicates, the ones that make siliceous rocks. And because this means most rocks, apart from limestone and the other calcareous materials, they are often referred to as the "rock-forming minerals." Let's be clear at the outset that with these siliceous rocks we're talking about *silicate* compounds, which involve the element *silicon* and a subgroup of *silica* minerals, which we'll come to at the end of the chapter. None of this has anything to do with *silicone*, the synthetic polymer of multifarious uses.

The principles discussed here are the same as those developed in the previous chapter, but the silicates present special challenges. Indeed, for a long time they were very tricky things to understand at all. The early geologists had at their disposal new ways of chemically analyzing minerals, and they applied them with gusto. They made impressively rapid progress, but they were baffled by the silicates. Their analyses showed that these minerals were dominated by silicon and oxygen but beyond that, well, they seemed too numerous, wildly varied, and inconsistent. It turned out to be well into the twentieth century before there was a breakthrough in understanding these perplexing compounds.

The breakthrough took place when it dawned that X-rays could be used to study the structure of crystals—any crystals: geological, metallurgical, biological, and so on. Today, over twenty Nobel prizes have been awarded for work in this field, most recently in 2012. It's even relevant to wine itself, through elucidating the structure of enzymes, proteins, and the like. (Incidentally, the use of X-rays to analyze crystals is quite different from its use in producing the familiar X-ray pictures of the human body.)

In July 1912, the Bragg family rented a house, Whin Brow, at Cloughton, high above England's Yorkshire coast. The father of the family, William, was a Professor of Physics at Leeds, and his 22-year-old son, Lawrence, was a precocious physics student at Cambridge. With an impending war, they were eager to escape the grimness of city life for a while, but even so, William took some work with him. And it was while puzzling over some new X-ray data that the two vacationers suddenly saw a way of calculating crystal structures. Some vacation! The following year, Bragg father and son successfully determined a complete crystal structure for the first time, that of sodium chloride, and they then went on to tackle a range of other crystalline substances. Along with other workers, they made stabs at the silicates, but these complex compounds remained inscrutable. It was 1926 before the Bragg team finally established the crystal structure of a silicate (the mineral olivine) and at last presented a solution to the silicate riddle. In July 2013, a plaque was placed on the Whin Brow house to commemorate that epochal vacation, one that initiated the understanding of a host of compounds, including cholesterol, insulin, hemoglobin, vitamins, antibiotics, and DNA, as well as the minerals that make rocks!

The key to the silicate minerals that the Braggs discovered is that all are based on particular building blocks of silicon and oxygen, linked together to different degrees (see the accompanying box). How this concept works will become clearer as we turn to examples of the minerals, arranged in order of increased linkage between these building blocks. At first reading, it may seem complicated, but as William Bragg himself remarked, the scheme has an order and an elegance that the early geologists could only have dreamed of.

The Secrets of the Rock-Forming Minerals

Three principles together unlocked the secrets of these minerals:

1. *All the silicate minerals are assembled from a basic "building block" that consists of a silicon ion surrounded by four oxygen ions in the form of a four-cornered pyramid. This little jewel is the key to it all, so we will call it by its proper name: a* **tetrahedron,** *plural* **tetrahedra** *(Figure 3.1).*
2. *The silicate tetrahedra can link themselves together to various degrees by linking corners and sharing the oxygen at the mutual corner. The extent of this linking and sharing provides the basis of a subclassification. For example, in the following pages we will meet silicate subgroups that lack any sharing, some that have adjacent tetrahedra sharing two corners to give long chains, and others that have three of the four tetrahedral corners linked to a neighbor to give a sheet-like arrangement. This latter structure is the basis of the clay minerals, which are so important to viticulture.*
3. *Combining one silicon and four oxygens into the building block, the tetrahedral silicate ion, leaves it with two negative charges, SiO_4^{2-}. These negative charges have to be balanced by cations in order to give overall electrical neutrality, but the number required will depend on the extent of tetrahedral linkage. As more tetrahedra are linked and more oxygens are shared, the negative charge decreases. Hence, different cations are appropriate for differing amounts of linking. For instance, one of the most common elements in vineyard soils is iron,*

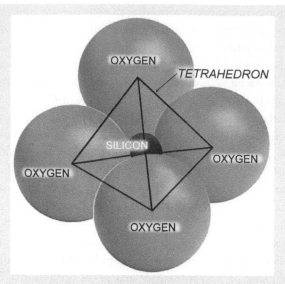

Figure 3.1 The silicate tetrahedron, showing how the smaller silicon ion sits at the center of a pyramid-like arrangement of oxygen ions at each of the four corners.

and feldspar is one of the most widespread minerals; yet, the two are normally not found together. There is no iron-bearing feldspar mineral. This is because in feldspar, all four corners of the neighboring tetrahedra are linked to give a three-dimensional mesh structure, and the size and electrical charge of iron make it unable to fit in.

Olive Green and Garnet Red

In some silicate minerals, the tetrahedral building blocks are *independent*; there's no sharing at all. Here, each tetrahedron carries two negative charges and is bonded to its neighbors simply by cations fitting between them and accepting electrons. Several different cations can do the job, but the most common ones are iron and magnesium. The bonding can be carried out entirely by iron, by magnesium, or by any combination of the two, rather like the calcium–magnesium gradation we met in Chapter 2.

Early geologists tended to give names to each of these various permutations, regarding them as separate minerals, and this obviously led to a great proliferation of terms. We can simply call an iron–magnesium silicate mineral with unlinked tetrahedra, **olivine**. Just how much or how little iron it contains will have depended on the chemical environment at the time of its formation. Generally, olivine has some intermediate amount of iron—say, 20 or 30% of the cations—with the complementary 70 or 80% in these two examples being magnesium.

One reason for mentioning this chemical complexity here, and it applies to many of the following silicate subgroups, is that this cation variation can affect some of the physical properties used to identify the mineral, including the most immediate one: color. Most olivine has a yellow-green olive color, hence its name, but this color darkens with increasing iron content to the point that pure iron olivine becomes almost black. Next to Lake Balaton, Hungary, the relatively iron-rich olivine makes the soils very dark, almost black, which reportedly helps them retain heat and aid grape ripening.

Because the relatively small ions of iron and magnesium bond the silicate tetrahedra into an efficient three-dimensional meshwork, olivine minerals tend to be hard and dense, and to lack planes of mineral cleavage, looking rather like dark-green glass. They can be an important constituent of volcanic rocks and soils, as in, for example, the Macedon Ranges of Victoria, Australia, and the Kaiserstuhl of southwest Germany, especially at Limbert Mountain near Sasbach. Around La Geria on Lanzarote, nuggets of deep-green olivine glint in the ocean sun.

We may come across yet other minerals with unlinked silicate tetrahedra in vineyards and on wine labels. Among them are **epidote**, a calcium silicate typically seen as yellow-green (the textbooks say "pistachio green") shapeless masses in rocks and soils associated with metamorphic limestones, and **garnet**, a well-known mineral that can have various permutations of iron, magnesium, calcium, aluminum, and manganese. These chemical elements help bond the silicate tetrahedra together, giving a range of minerals with various names and colors. The most common is the iron–aluminum variety called **almandine**, with the dark wine-red color associated with garnet. Tasting notes on red wines may mention a garnet hue, obviously referring to almandine, and not to, say, the calcium-rich garnet called grossular because of its resemblance to a green gooseberry! Garnet is hard and resistant to weathering, and so it occurs in some vineyards as small rounded granules, as it does in eastern Corsica and around St. Fiacre in Muscadet, France.

Dark Horses: Pyroxenes and Amphiboles

In this diverse group of minerals, many of which are dark in color, two corners of a silicate tetrahedron are shared with its neighbors to give a chain-like arrangement (Figure 3.2). So in these *chain silicates*, any particular tetrahedron has two of its corners linked and the other two corners unlinked. This sharing reduces the negative electrical charge of the silicates, and fewer cations are needed to provide electrical neutrality. Again, iron and magnesium fit the bill, but so do a number of other cations, particularly calcium, giving a wide variety of appearances and a host of mineral names. It's worth introducing here a term that will be useful when we come to look at the igneous and metamorphic rocks: **mafic** minerals are those rich in iron and magnesium, such as olivine and the minerals discussed here. In general, they are

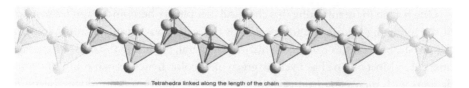

Tetrahedra linked along the length of the chain

Figure 3.2 Diagrammatic representation of silicate tetrahedra with two corners linked lengthwise to form a chain, as in the pyroxene minerals. The tetrahedra are shown "exploded" slightly, with the ions separated to show the chain structure more clearly.

dark in color and so are found in generally dark rocks such as basalt and gabbro (Chapter 4) and amphibolite (Chapter 5).

All minerals in this group have a more or less elongate shape, rod or stick-like, exactly in line with the elongate arrangement of the chains of silicate tetrahedra down at the atomic scale. All these minerals have two sets of cleavage planes running along the length of the mineral grain, but these planes are hardly visible without magnification. The cleavage planes make parallel hairlines along the mineral length. When viewed end on, so to speak, the two different cleavage sets may be discernible, giving two arrays at a high angle to one another (Figures 2.2 and 2.3). In other ways, however, these minerals have variable appearances. Color is a case in point: they can be blue or green; pure calcium examples are white; iron-rich varieties are black.

We can further subdivide these chain minerals into two families, **pyroxenes** and **amphiboles**, according to how the chain structures are organized. The most common pyroxene is the dark green, iron magnesium calcium mineral called **augite** (though probably the best known is jade). The most widespread amphibole is the complex, green-black, highly elongate variety known as **hornblende**. Both of these minerals help form a variety of igneous and metamorphic rocks. Examples are found in the vineyard soils of Heathcote (Victoria, Australia) and Vesuvius in Italy. Reportedly, Barbera di Colli Tortonesi from the Piemonte region of Italy is distinctive partly because of a particular pyroxene in the soil there. Because these minerals are reasonably widespread around the world, various of their names have found their way onto wine labels and vineyard designations.

Minerals in Sheets: Mica, and so on

We come now to the *sheet silicates*, perhaps the most complex minerals of all but arguably the most important for us. This group includes the fabled clay minerals, which are so influential in vineyards both chemically in underpinning nutrient availability and physically in controlling drainage. We will deal with the clays in their own section later in this chapter. The basic architecture of these sheet minerals

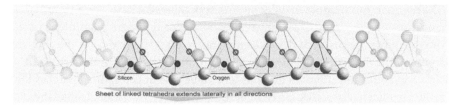

Figure 3.3 Diagrammatic representation of silicate tetrahedra linked laterally to form a sheet, as in the micas and clay minerals. The sheet extends (effectively, infinitely at this molecular scale) sideways, away from, and out toward the viewer. One corner of each tetrahedron, in this instance those pointing upward, is left unlinked.

is straightforward enough: each silicate tetrahedron shares three of its corners with neighbors to form a continuous leaf (Figure 3.3), and various cations bond adjacent sheets into stacks. But things are in reality much more complex than that: the tetrahedra linked within a leaf can have differing arrangements, some with the remaining free corner pointing upward and others pointing down. These differing sheets can be stacked in all sorts of permutations, and their differing separations allow a wide range of cations to do the bonding. It is hardly surprising, then, that the sheet silicates are so very diverse.

Even **mica** is actually the name for a whole family of minerals, but they're all distinctively platy in shape, directly reflecting their sheet-like structure down at the atomic scale. Some early geologists gave these minerals the lovely name of "glimmer." Only two of these "glimmerous substances" need concern us here. The first is **muscovite** mica, a colorless, potassium aluminum silicate with no water involved. It's so named because large translucent sheets of it were once used as windowpanes in Moscow. It's soft and distinctively sheet-like, looking almost like films of perspex. Many a beginning student has looked at a specimen box labeled muscovite and thought it empty, containing only a lining sheet of clear plastic!

Muscovite is used in the little viewing windows of old-fashioned oil stoves and the like, sometimes still being called by its old name isinglass. (There are folk who love debating how the isinglass curtains in the musical Oklahoma's "Surrey with the Fringe on Top" could possibly "roll right down" if they were rigid leaves of muscovite!) The flat, shiny surfaces of muscovite flakes strongly reflect sunlight, giving the glittering look of some micaceous soils, say at Poncie, in Fleurie, France.

Looking very similar in most ways but distinctively dark colored, even black, is the other common mica: **biotite**. No aluminum is involved here; the sheets of silicate tetrahedra are bonded together by iron and magnesium. Biotite occurs, for example, in some Alsace vineyards, such as the Grand Cru Brand, just outside Turckheim, Alsace, and in Pewsey Vale, South Australia. A Vinho Verde (Portugal) is actually labeled "Biotite," after the vineyard soil.

Micas occur in a number of different rock types; muscovite and biotite are common in granite (in some cases together) and typically as easily visible crystals. Consequently, vineyard soils derived from granite typically sparkle in the sunlight because of the myriad sheets of mica. The Junrode vineyard of Condrieu, France, is a case in point, and although the soils tend to be pale in color, those patches that overlie a granite bedrock that is rich in biotite are noticeably darker. Unlike the clay minerals, micas have a negligible CEC and don't absorb water. Consequently, micaceous soils don't participate in any direct way with nutrition, and they tend neither to store water nor to lose their drainage properties even when wet. In other words, micaceous soils are generally inert and very well drained.

The sheet silicate called **glauconite** is found in some sedimentary rocks, particularly sandstones, and it has two aspects that are of interest to us. First is its chemistry. A whole array of different cations can be involved, but potassium and iron dominate. In fact, glauconite is mined as a source of potassium fertilizer. Promotional material from places such as La Chautagne in Savoie, France, the Vipava area of western Slovenia, and the coastal districts of McLaren Vale, Australia, like to mention the beneficial presence of glauconite in the soils there. Because of the iron content, soils that contain glauconite tend to avoid chlorosis; growers in the Chinon-Saumur district of the Loire Valley have long been aware of chlorotic yellowing leaves on the region's calcareous soils but their glauconitic sandier patches yield nice green foliage.

The second relevant aspect of glauconite is its color, but this leads to some confusion. Glauconite tends to be greenish, and it has been much used as a pigment, even in ancient times: it's found in prehistoric cave paintings around the world and in Native American artwork; painters of Russian icons also favored it; Vermeer used it to give pallor to skin tones. Hence, small amounts of glauconite can give rocks a green color, from which comes the term *greensand*, which a number of wineries in New Zealand and England use very fondly.

A green sand (note the space in the term) could be colored by any green mineral, such as chlorite (below) or olivine, but greensand (all one word) implies by convention that the green color is due to the mineral glauconite. In a subtle extension of this terminology, a whole collection of greensand layers in the bedrock of a particular area with just minor amounts of other rocks (what geologists would call a formation) is labeled with a local name. Here the term is capitalized. Thus, New Zealand, for instance, has the Waihao Greensand, the Kokoamu Greensand, and the Gee Greensand. This usage is particularly important in southeast England where almost all the wineries are located either on the formation known as Chalk (also capitalized when used in this way) or on Greensand. As it happens, both offer a good balance of drainage and water storage, but the glauconite of greensand can also provide the iron and potassium that chalk may lack.

The mineral **serpentine** has only magnesium linking the silicate sheets together, and water is bonded inside the structure. It's the result of water reacting with magnesium-rich olivine and is the basis of the metamorphic rock serpentinite (Chapter 6). This preponderance of magnesium can produce soils that can be severely imbalanced, with associated minerals rich in nickel and chromium exacerbating the effect. Hence, soils rich in serpentine can be that unusual thing, actually inimical to grapevines.

Chemically very similar to serpentine is the mineral **talc**. It occurs in metamorphic rocks that are rich in magnesium, but it is probably of limited distribution in vineyard soils. It is mentioned here because the word "talc" appears in tasting notes. Like all silicate minerals, talc is flavorless—comprehensive data sheets mention neither taste nor smell—but the noteworthy aspect of talc is its softness. The silky feel of powdered talc is well known; larger pieces of it feel distinctly greasy, and a fingernail easily gouges it. Therefore, tasting notes presumably relate the mouthfeel of a wine to this exceptional softness.

Chlorite is the name for a group of dark green minerals with sheet silicate structure, involving mainly magnesium, iron, nickel, and manganese. The name has nothing to do with various chemicals called something-chlorite, which are occasionally used in wineries as bleaches and sanitation agents. It's an important constituent of rocks such as slate and schist, and it tends to give them a green tinge. Examples of vineyard soils containing appreciable amounts of chlorite are in northeast Corsica, Banyuls and the Durban area of Corbières (France), and at Kastelberg, on the outskirts of Andlau, Alsace.

Tiny Minerals with Huge Effects: The Clay Minerals

And so we reach the sheet silicates known as **clay minerals**. Incidentally, we use the term *clay* for both the very finest sediment and the material of which it is usually composed, the clay minerals we are discussing here. The clay minerals are an especially challenging group in that they are very complex, difficult to work with, and in many ways far beyond the scope of this book. But they are paramount in the vineyard, and so we have to take a closer look at this mineral group, though being very selective.

Practical difficulties arise both from the sheer complexities of the sheet structures in clays and from the diminutiveness of the individual flakes of the minerals. Most definitions place the maximum size at 0.002 millimeters (two thousandths of a millimeter), which means that advanced laboratory instruments are required to even see individual clay particles. We have already met the far-reaching consequences of the surface area this provides, but the figures are staggering. A handful of clay will contain a surface area more than a thousand times greater than a handful of sand. Just a single gram of dry clay can include a surface area greater than a couple of parking spaces.

Other important properties of clay minerals arise from their very nature: their sheet structure. Some of these minerals expand upon heating, and others allow water to penetrate between their constituent sheets and to be stored there; such a feature can be critical for a clayey soil in an arid ripening season. At the same time, this incursion leads to the expansion of the clays so that they start to clog drainage channels in the soil, in wetter times leading to waterlogging. So it's a delicate balance between water storage and drainage, which, together with the effects on fertility, makes clays fundamental to the behavior of vineyard soils.

Kaolinite

The mineral **kaolinite** is one of several chemically different minerals known together as kaolin. In many ways, it's the simplest—well, the least difficult—clay mineral. As clays go, each flake is large, around five times the size of illite and 250 times larger than montmorillonite (see later), and it has a relatively simple composition, with just aluminum and hydroxyl (OH⁻) ions in addition to the silicon and oxygen. The resulting bonding is strong, with none of the complications that arise in other clays that involve other metal ions. Thus, the CEC of kaolinite is low, and it has little tendency to swell when wetted, so clayey vineyard soils can drain relatively well if the mineral is kaolinite.

Like other clays, a mass of moist kaolinite can be molded, but the tight bonding of the sheets means that it can form strong, thin slabs—hence its widespread use in pottery. Its name comes from the hill where the mineral was first reported—some say by Marco Polo—Kao-ling, near Jauchau Fu, in Jiangxi Province, southeast China. So famous were these deposits for pottery that the product became known simply as "china."

Kaolinite forms in a variety of circumstances, but in vineyards it's most commonly the weathering product of the feldspar and muscovite in aluminum-rich rocks like rhyolite and granite. In parts of Sonoma County, California, kaolinite soils weathering from rhyolites are preferred because of their good drainage and, because of the low CEC, the restricted vigor of the vines. At the same time, kaolinite can absorb and store a little water. At Fish Hoek, on the Cape Peninsula of South Africa, which has a working kaolinite mine, some growers believe that the granite soils would be excessively drained were it not for their high kaolinite content. Similarly, although its absorption properties are much less than those of some other clays, kaolinite is used to some extent as a wine fining agent. It is also marketed as a vineyard spray, to act as both a sunscreen in exceptionally high-sunshine areas where leaves can become scorched (e.g., Queensland, Australia) and as a physical barrier to vine pests, a technique allowed in organic viticulture.

Kaolinite is also found in recent volcanic tephra, weathered from the material known as **allophane**, itself formed from the breakdown of volcanic glass. Allophane is a curious substance, being rather like kaolinite in its chemistry but with a very poor internal organization—barely crystalline. It occurs as fluffy balls of gel, sometimes hollow, which are similar to silica gel in their ability to hold water, as well as attracting humus. Some growers in places like the southern flanks of Etna (Italy) and Tenerife (Canary Islands, Spain) consider allophane an important component of their vineyard soils.

Smectite/Montmorillonite

The names "smectite" and "montmorillonite" are blurry. Both appear in vineyard literature, sometimes interchangeably, sometimes with smectite as a member of a montmorillonite group, and sometimes treated as two different minerals. However, in line with much modern usage, I will use the term **smectite** for a group of minerals, an important member of which is the mineral montmorillonite.

Montmorillonite is particularly fine-grained and can involve a range of cations, giving it, along with the other smectites, a high CEC. It's named after the town of Montmorillon, in the Poitou-Charentes region of western France. An important characteristic of all the smectites is their behavior with water. Substantial amounts of water can be adsorbed both onto the surfaces of the clay flakes and between their internal sheets, leading to significant expansion. In fact, sodium-rich montmorillonite can swell to several times its original size (thereby, incidentally, increasing its cation exchange capacity). Hence, the smectites are sometimes called "swelling clays."

As mentioned earlier, the storage of moisture provides obvious advantages in dry viticultural areas with well-drained soils. However, swollen clays at the ground surface may make it curb any further water penetration, and at depth they may impair penetration by the finer, water-seeking root hairs. Also, in times of drought, the opposite process occurs: the clays give up their internal water and shrink! This phenomenon of shrink-and-swell can be problematic, not just for vine roots but also for building foundations. (In the United Kingdom, where such clays are widespread and often called heave clays, they account for the single biggest year-on-year insurance claim for building damages.)

Most smectites originate from the weathering of igneous rocks such as basalt and various volcanic deposits. For example, the young volcanic rocks of the Tokaj-Hegyalja region, Hungary, are currently weathering into smectite minerals which, in addition to helping form some of the best vineyard sites, are being quarried for industrial use. But geologic processes can rework such volcanically derived material and re-deposit it as sediments. Hence, many areas of sedimentary rock involve soils rich in montmorillonite and other smectite clays. Because limestone is notorious for poor water-holding, such soils can come to the rescue in arid vineyards. In the area around Cognac, France, for example, the bedrock is a very dry limestone, but sufficient water for the vines is usually held by the overlying montmorillonite-rich soils. Some assert that the specialness of the soils of Chateau Petrus in Pomerol, as well as some of the Grand Cru sites at Vosne, Burgundy, is due to the intricate interplay between the local climate and the high smectite content of the soils.

These kinds of clay used to be known as *Fuller's Earth*—and in some quarters still are—reflecting their ability to degrease and thicken materials such as wool ("full-ing" is an old word for treating cloth). But as far back as 1811, John Pinkerton wrote in his book "*Petralogy (sic): a Treatise on Rocks*" that the term *Fuller's Earth* "would seem to be rather a solecism." Instead, he suggested a term based on *smēktis*, "the equivalent Greek denomination," namely, smectite, and it is this term that has stuck in geology.

Montmorillonite is also widely referred to as **bentonite**, the fining agent that is much used in wineries. Mixed into a slurry and added to wine, the swelling sheets are able to absorb potentially awkward proteins and help ensure that the wine is clarified and remains so. Bentonite is also sold as a soil conditioner, used to increase the water-holding abilities of soils and to give them weight. The swollen material has the effect of "thickening" the soil, making it "heavier" (Chapter 10)—and stickier. Anyone who has tramped a vineyard with smectite soils—Vaillons in Chablis, say—in the rain will know how eagerly the sticky clumps cling to your boots. (Just as in the hell that was Flanders during World War 1; 90% of the Ypres clay is smectite.)

Geologically, most commercial bentonite is around 80% montmorillonite, the remainder being some mix of other smectites and silicate minerals. Incidentally, the material is named for Fort Benton, which is often quoted as being in Wyoming. Indeed, there is a (ghost) town called Benton in Wyoming, and much bentonite today is mined in Wyoming, from a formation called the Fort Benton shale. Fort Benton itself, however, is over the state line in Montana.

Illite

Illite is a potassium aluminum clay named for the state of Illinois, where in 1937 it was first recognized. It has a lot of similarities with the potassium-bearing mica muscovite, except that it is clay-sized and that water is involved in the structure. Also, some calcium and magnesium can substitute for the potassium, and these firmly bond the silicate sheets together such that outside water cannot easily enter the structure: illite neither swells with water nor shows much tendency to expand on heating. The substitution does lead to some CEC, but the potassium is firmly bonded in the crystal lattice and is not easily released. Growers generally treat illite clay merely as a background, long-term source of potassium.

Illite forms through the weathering of potassium and aluminum-rich rocks like granite and is itself fairly stable. Thus, it survives transport in rivers and out into the sea, such that around half of the clay deposits on the seafloor consist of illite. In turn, many sedimentary rocks that form from these sea-bottom sediments contain illite, such as shales and clay mudstones. Illite has been reported, for example, in the vineyards of the Finger Lakes region in New York State where it is derived from the shale bedrock.

Vermiculite

Vermiculite results from the weathering of micas such as biotite, chlorite, and magnesium-bearing amphiboles, the kinds of minerals found in volcanic rocks. Its crystal structure allows its silicon to be replaced by both aluminum and magnesium (with just half the charge of silicon), a charge imbalance that leads to a particularly high CEC, greater even than montmorillonite.

Like most of the clay minerals, vermiculite contains water. But unlike the loosely attached liquid water that participates in the kinds of exchange and swelling processes outlined above, in vermiculite this water is bonded wholly within the solid crystal lattice: its chemical formula is written to include H_2O. However, if the mineral is heated very quickly, this water can be driven off—turning to steam and blowing apart some of the mineral layers into an expanded, accordion-like form, likened to worms in appearance and hence its name. The kinds of temperatures usually employed—upward of 100°C—mean that the process is irrelevant to the vineyard itself, but all growers know that this material is useful for raising plant cuttings. Having been expanded, its low density gives easy handling and good drainage, while maintaining its excellent CEC. Thus, clay minerals are important not only in the vineyard, but also in the world of vine nurseries. And, for that matter, vermiculite is useful in bricks, pots, packaging, brake linings, loft insulation, explosives, cat litter....

The Workhorses: Feldspar and Quartz

This final group of silicates includes the two most abundant minerals in the Earth's crust: feldspar and quartz. Both are *framework silicates*. Here each silicate tetrahedron shares all four of its corners with neighboring tetrahedra (Figure 3.4). However, in the feldspar minerals, aluminum substitutes for some of the central silicon ions, and this leads to electrical charges that cations have to neutralize. The cations with appropriate negative charge and the right size to fit the restricted space within the framework structure are nearly always potassium, sodium, and calcium. The iron and magnesium that are so important in many of the silicate minerals are unsuitable for the job, and hence these are not mafic minerals. The complementary equivalent, as typified by feldspar, is **felsic**. These felsic minerals tend to be light in color, so, along with quartz, the rocks they form are generally pale. Granite, sandstone, and quartzite are examples.

Feldspar is actually the name for a group of related minerals. At one time, the names of all sorts of minerals, even icicles and hailstones, had the suffix "spar" appended to them, but the only one that officially held out is this "mineral of the field." The two main representatives of feldspar are plagioclase, which can be rich in either sodium or calcium, and potassium feldspar. Even so, it is

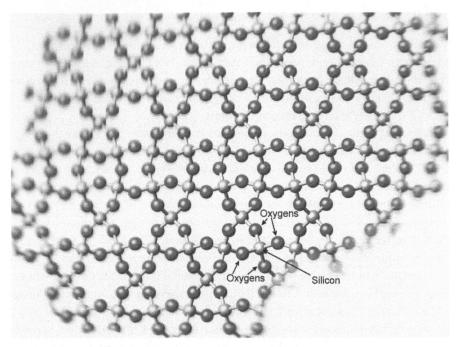

Figure 3.4 Diagrammatic representation of a slice through a three-dimensional framework of silicate tetrahedra with all four corners linked, as in the mineral quartz. You have to imagine the meshwork extending toward you out of the picture, as well as backwards and in all the other directions.

often handy to simply refer to both of these variations simply as "feldspar." The meshwork of silicate tetrahedra in **plagioclase** is such that in two directions the linking is systematically less strong, and consequently the mineral has two directions of cleavage. They aren't quite at right angles to each other, and this is the origin of the mineral name, *plagio* being the Greek for oblique. The "clase" part of the name, which means breaking, comes from the same root as the *clast* in volcaniclastic rocks.

The chemical situation with the sodium and calcium in plagioclase is exactly analogous to the one we have explained for magnesium and iron in olivine. In general, plagioclase is a pale-colored mineral because the sodium-rich varieties are pale gray (white if the cation is entirely sodium), which are more common. The calcium-rich varieties are darker, with the purely calcium variety being almost black. A glance at Figure 4.2 shows the importance of these plagioclase minerals for the igneous rocks and in turn other rocks derived from them.

The weathering of plagioclase leads to clay minerals that can be a significant source of calcium for vines. Studies in the Eden Valley of South Australia documented a systematic decrease in grape sugar levels but increased acidity with increasing levels of calcium in the soils, most of which derived from changes in the bedrock plagioclase. Appreciable amounts of plagioclase in vineyard soils have also been recorded in the Sveta Lucija district of Istria (Croatia) and the basaltic soils of the Walla Walla Valley, Washington.

Potassium feldspar is pale colored, usually gray or pink, and in soils it tends to make box-shaped pieces, such as below the Helderberg in South Africa (Figure 3.5; see Plate 4), in Dão (Portugal), or at Temecula (California). It accounts for the pink color of some of the Beaujolais soils, for example, around La Madone, in Fleurie. Over time, some of the potassium is weathered out of the feldspar, possibly then being leached away. Otherwise it leaves what remains to reorganize and becomes incorporated into kaolinite, the dominant weathering residue. Conventional wisdom holds that granite soils such as those in Beaujolais are unsuitable for growing Pinot Noir, even though it does so well not far to the north but there in calcareous soils. Despite that dictum, it thrives in Alsace in soils derived from granite and rich in potassium feldspar, such as on Gloeckelberg near the village of Rodern—a place some call "the cradle of Pinot Noir."

Now to the **silica** minerals: these minerals have the most straightforward chemistry of all the silicates, being just silicon dioxide, SiO_2, silica. **Quartz** is made of silica and must be one of the best-known minerals of all, if only as a name. The name

Figure 3.5 Box-shaped fragments of potassium feldspar in the soils of the Uva Mira wine estate, Helderberg, South Africa.

pops up in a surprising array of contexts (watches, videogames, kitchen worktops, etc.) and not least in the names of vineyards and wines as well as in tasting notes. There are two reasons for this familiarity. It's not only one of the two most common minerals in the Earth's crust (forming about 12%), but being tough, durable, and eye-catching, it is often noticed. Incidentally, like the term *sheep*, *quartz* covers both the singular and the plural; geologists don't talk of "quartzes."

We have here the ultimate linkage state of silicate tetrahedra, giving a relatively dense packing of the silicon and oxygen atoms, and therefore good hardness and no mineral cleavage. It breaks with a **conchoidal** fracture (Chapter 2), just like glass. Quartz occurs in many kinds of rocks—igneous (especially granite), sedimentary (especially sandstone) and metamorphic (especially the rock quartzite)—typically as glassy-looking small grains. Its most typical appearance in vineyard soils is transparent to milky white, glassy, roundish granules. Quartz is a common vein-forming mineral in rocks (Chapter 7), so irregular blebs and zones of milky-white material crossing bedrock may well be made of quartz. Because of its resistance to weathering, quartz is also a common constituent of sediments, such as screes, river gravel, and beach sand.

This lack of complication, however, produces a paradox: although pure quartz is colorless, the simple chemical structure and composition allows just tiny amounts of impurities to produce strikingly colored variants. Less than 0.1% of iron in the silica gives **amethyst**; trace amounts of aluminum replace silicon in **smoky quartz**; **aventurine** has minuscule amounts of chromium; tiny variations in chemical impurities between roughly concentric bands produce the well-known appearance of **agate**. White quartz with a pronounced opaque look is widespread and is sometimes referred to as **milky quartz**. It's due to infinitesimal bubbles scattered throughout the crystal lattice.

If the mineral is pure enough to be colorless and clear and was able to form with regular, external crystal faces (Figure 2.1) it's often referred to as **rock crystal**, and if it is able to approach its ideal external shape, there may well be a six-sided pattern. This shimmering, symmetrical appearance, transparent like glass or ice (Pliny thought that quartz was a permanent form of ice), is the archetypal crystal of the popular image, and, to some, it engenders an otherworldliness. And presumably it's the image that underlies phrases involving "crystalline" in tasting notes (albeit inconsistently, with some using the word to describe appearance or limpidity, some to indicate purity of flavor, others to describe sharpness, and yet others, sweetness).

Silica occurs in further forms that are relevant to viticulture and wine. The relative simplicity of the silicate framework enables it to form in extremely tiny crystals, needing a very powerful microscope to make them out. The two frequently encountered names are **flint** and **chert**. There is some debate in geology about the precise usage of these terms; I will treat them here as synonymous. Flint (and therefore chert as well) has a conchoidal fracture like quartz, but the surfaces are not as uneven and curving, making it easier to produce a fairly straight edge. This is exactly why flint was so favored for toolmaking before the advent of metalwork.

Flint is a dense, opaque to slightly translucent material, and typically varies from white to dark gray in color, sometimes within a single piece. Thin chips tend to show greater translucence at the edges. It's tough and inert, which is why flint is tasteless and odorless, despite its common use as a metaphor in wine tasting (Chapter 12). Flint occurs in sedimentary rocks in smooth but highly irregular, bulbous shapes termed by some geologists **nodules**. Others refer to such bodies as **concretions** (Chapter 5). Again, the naming is inexact. Most flint nodules formed while the host sediment was just beginning to harden, due to the breakdown and local reprecipitation of marine organisms made of silica, such as certain kinds of plankton, sponges, or the tiny organisms called radiolaria. The word **silex** also appears in wine writings in English, perhaps adding a touch of the exotic, although it's simply the French word for flint.

Jasper is an unusually opaque form of finely cystalline silica, often with an intense color, most commonly blood red. In places such as the Frankenwald (Upper Franconia) and Vogtland areas of eastern Germany, very fine carbonaceous matter gives it a dense black look referred to as **lydian stone** or **lydite**. The name comes from the ancient province of Lydia, in Asia Minor, but it is probably best known in the wine world as a constituent of pebbles in some Médoc gravels, especially around Chateau Palmer. As long ago as 3500 B.C., Lydian stone was used to make tablets on which lines were lightly drawn with soft metals, such as alloys of gold. The traces differed in color according to the gold content of the alloy, and so the purity of gold could be judged. That is, lydite was the *touchstone*.

Finally, after all these permutations of silicate tetrahedra and crystal lattices, there's **opal.** Here the silicate tetrahedra allow water to come in between them, and they become wholly disorganized. It's this lack of any real pattern in the arrangement of the tetrahedra that gives the substance its characteristic shimmer. In other words—although it may be a strange ending to a chapter dealing with intricate crystal structures—opal is that rare and anomalous geological material: a mineral that is not actually crystalline.

Further Reading

The kinds of highly illustrated handbooks on minerals mentioned in Chapter 2 include the silicate minerals and their appearance. More technical aspects, such as silicate tetrahedra and clay minerals, require books at a more advanced level. Here, the standard academic work is W. A. Deer, R. A. Howie, and J. Zussman, *An Introduction to the Rock Forming Minerals* (3rd ed.). London, Mineralogical Society of Great Britain & Ireland, 2013.

4

Igneous Rocks

Molten Rocks Beneath Our Feet

Igneous rocks were once molten. This is a simple statement, but it's exactly what sets them apart from the other two great divisions of rocks: sedimentary and metamorphic. So, deriving from the Latin word for fire—*ignis*, the same word that gives us *ignition*—igneous rocks are associated with heat. Some are simply solidified lava, but most originated by slowly cooling below the Earth's surface. Thanks to erosion through time of the overlying material, such rocks are now widespread at the Earth's surface and consequently underlie many of the world's vineyard regions, from Washington State to the mountains of Hungary, from Lodi, California, to the Cape Peninsula of South Africa.

Although it gets warmer with depth everywhere across the Earth, generally the weight of the overlying rocks makes the pressure too great to allow melting, so as a rule the rocks below our feet are solid. In some places, however, the heat increases so rapidly that temperatures can reach over 600°C at just a few kilometers below the ground surface, a temperature at which some rocks are molten, even under pressure. The initial melting usually takes place in and below the lower part of the Earth's crust, but the molten rock then rises, typically to reside tens of kilometers or so below the surface, though less under volcanically active areas. Such depths may seem large to us, but seeing as its well over 6000 kilometers to the center of the Earth, geologically they are pretty close to the surface. In other words, the igneous rocks we now see at the surface did not form incredibly deep in the Earth's interior; they were nowhere near Earth's core, as some writings claim.

We call this underground molten material **magma**. People seem to like the word. Not only does it appear on wine labels, but it is also the name of a number of wine shops, bistros, and various drinks. It exists in the Earth in **magma chambers**. It would be simplistic to picture these as some sort of enormous underground caves filled with liquid rock: there may be patches that are wholly liquid, but almost certainly there will be plenty of solid matter, minerals that are below their melting point. So we often talk of a "mineral mush" for this partly molten, partly solid mix, and the magma is likely to be increasingly mushy toward the chamber margins, where it grades into wholly solid rock.

There are two ways of thinking about igneous rocks. One is to look at what they are made of, that is, their constituent minerals and by implication their chemical composition. This approach is the basis of classifying igneous rocks into groups such as granite and basalt, and we will consider these names later in the chapter. The other approach concerns the situation in which the molten material came to become solidified. Did it congeal underground or at the Earth's surface, having been erupted as a volcanic lava? It's a basic distinction, and a closer consideration of this aspect leads to a better understanding of how igneous rocks work. So this is where we begin.

Below the Ground and at the Surface: Intrusive and Extrusive Rocks

Magma is less dense than the solid rock that surrounds it, and so it will try to rise upward. But, of course, the rock above is in the way! The mass of magma may be sufficiently hot to melt the rock in its upward path, but it will encounter ever cooler rocks on its way upward. It will progressively lose heat, and there may come a point where it is thoroughly solidified and can rise no further. A rock that is formed in this way—the result of invading other rocks and solidifying below the surface—is called an **intrusive** igneous rock.

A rising magma will exploit any fissures in the rock that roofs the chamber, and it may have sufficient internal pressure to generate some new fractures and to force rocks aside (Figure 4.1). Blocks of the solid host rock may break off and fall into the magma: we call these **xenoliths**. The boulders of granite that dot the vineyards in Amador County, California, in the Sierra Foothills AVA, contain many examples. Up in the Sierra Nevada from where they were derived, many bare rock faces (in Yosemite, say) present striking examples, typically speckled, dark gray patches in the pale host granite. Roughly vertical fissures in-filled by the rising magma solidify to form a feature called an igneous **dike**. If the melt was injected along cracks in a more horizontal direction, typically by splitting the layers in sedimentary rocks, the structure is called an igneous **sill**.

Why do such things matter, apart from spurring academic curiosity about the world deep below our feet? Although these features are forming deep below the surface, with time and with the Earth's perpetual internal mobility, they may at some later stage be forced upward, and, coupled with the erosion that continually scours the surface, they may become exposed to be visible in the world around us. Thus, some of the vineyards on the Blue Ridge, Virginia, have outcrops where dikes are visible, and a sill is famous locally in the Broke Fordwich district of the lower Hunter Valley, Australia.

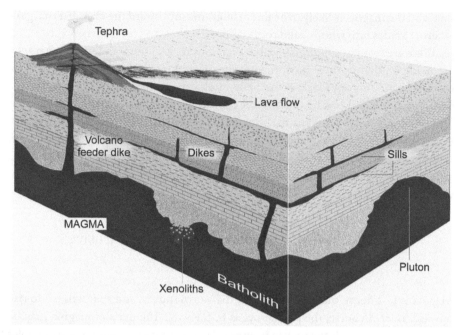

Figure 4.1 Some different forms of masses of igneous rock.

This natural excavation over time can "unroof" entire magma chambers that solidified more or less in place. These can cover vast regions, with decisive effects on the Earth's surface. Such huge bodies of igneous rock are called **batholiths**. They usually consist not of one mass but a coalescence of numerous lesser intrusive bodies called **plutons**. From the Rockies of British Columbia all the way down to Patagonia, the mountains that parallel the Pacific coast are dominated by batholiths of igneous rock. These intrusions can be enormous. The granite of California's Sierra Nevada is one such batholith, and although it probably formed over 30 kilometers below the ground, today it makes up the high, rugged spine of California.

When you stand among the vines on the flat floor of the Elqui Valley in Chile, those towering peaks all around you are carved in a batholith that stretches over 800 kilometers from north to south. The alluvial debris that underlies so many of Chile's vineyards is derived from it, this so-called High Andean batholith, as is much of the debris down around Mendoza, Argentina. There is an experimental vineyard way up in the Peruvian Andes, near Machu Picchu, at nearly 3000 meters above sea level. The vineyard and its surroundings, including the iconic view of the ancient Inca citadel with its stone buildings clustered beneath the towering peaks of Huayna Picchu and Cerro Machu Picchu—in fact all the rock and building stone you can see—is granite. It's all part of another vast batholith, this one around 1000 kilometers in length.

The middle Douro River in Portugal cuts its way through a number of granite plutons, each one part of the major Beiras batholith. A suite of granite plutons accounts for the mountains surrounding Stellenbosch and Franschhoek, in the Western Cape of South Africa. They then stretch northwestward for nearly 200 kilometers, through the wine regions of Darling and Swartland to Saldanha, on the coast. The granite soils of the wineries around Armidale, New South Wales, are derived from part of the New England batholith, a complex of igneous masses that continues northward to the Granite Belt wineries of Queensland, a distance of over 250 kilometers.

In such huge masses of magma, the rate of cooling will be very slow, allowing plenty of time for elements to diffuse through the magma and join an appropriate site in a growing crystal. Consequently, the resulting rock will be coarse-grained, in contrast with the fine crystals of rapidly cooling magmas. More familiarly, the same principle applies when water freezes to make ice crystals. Slow freezing leads to large ice crystals. (It's why a cold snap gives the fine, powdery snow beloved of skiers, whereas less cold temperatures give the coarse, slushy snow that's good for snowballs. It's why preserving berries by freezing is best done on "fast freeze," so that the ice crystals are tiny and are less likely to puncture the flesh, which leads to the shape of the fruit being lost on thawing.)

Sometimes the rise of magma may be intermittent. A magma may cool slowly at depth, allowing coarse crystals to form, but more rapidly in the later stages, especially if it reaches the surface. This is the normal explanation for a **porphyry**, an igneous rock with two or more distinct grain sizes and much prized by some vine growers (see the next section).

Igneous rocks have a distinctive texture of tightly intergrown minerals, with virtually no spaces between them, giving them great strength and resistance to erosion. There is none of the stratification that characterizes sedimentary rocks, none of the foliation of metamorphic rocks. But this means that the intact rock gives little access to vine roots and that there is virtually no water storage. We refer to this featureless appearance in outcrops as the rock being **massive**. An important reason why growers along the Douro in Portugal tend to avoid the granite there is its massive nature. Roots can barely penetrate the bedrock, and, unlike the case for the nearby schists, there is little opportunity to store winter rainfall for the arid summer.

Of course, magma can in some circumstances ascend all the way to the Earth's surface, breaking out as **lava**. This takes us into **extrusive** rocks and the realms of volcanism. The lava may be accompanied by all kinds of broken solid material, which we will discuss shortly. Chapter 9 explains how basalt lava tends to weather relatively quickly to yield a good range of plant nutrients and hence fertile soils. This is why, although much of southern Italy, for example, is based on limestone and is rather barren, the fertile soils on and around the volcanoes are prized by farmers.

Naming Igneous Rocks

The rocky material that makes the Earth is not uniform. Consequently, the chemical composition of magma will vary according to its place and depth of origin, thus giving rise to a range of different igneous rocks. Early geologists had difficulty making sense of this variation. One problem was that, in their attempts to classify igneous rocks, they were keen on utilizing the new chemical techniques that had just become available, and in some ways this just clouded things. In particular, the analyses revealed that some igneous rocks were particularly rich in silica, whereas others had much less. So, in line with chemical thinking of the time, the early workers supposed that the magma must have contained differing amounts of something they called silicic acid. Thus, for those rocks with lots of silica (such as those with quartz), there must have been an excess of silicic acid, and they came to be called "acid" rocks. But for those without quartz, the magma must have had relatively little silicic acid, having been used up in crystallizing the other silicate minerals. Again in line with the usage in chemistry, such rocks were termed "basic." So igneous rocks, with about 60% or more silica in their chemical composition, became known as **acid**, and rocks with less than about 50% silica as **basic**.

These terms are widely seen in wine writings but often lead to misunderstanding. In science, it has turned out that this whole notion of silicic acid in magmas was misdirected, and so the acid–basic usage for rocks has been discarded. These days geologists look not so much at the silica content of the rock but the remaining part. In "acid" rocks, this tends to be dominated by minerals rich in potassium and sodium, the felsic minerals (Chapter 3), and in "basic" igneous rocks it is dominated by the mafic minerals. Hence, the materials are now termed either felsic or mafic.

These new terms are to be encouraged because the old acid and basic usage leads to confusion: *acid and basic when applied to rocks have a different meaning from when they are used for soils* (or, for that matter, wine). As it happens, felsic igneous rocks such as granite tend to weather to yield acidic soils, but so do some mafic rocks. It is the calcareous rocks that yield basic (alkaline) soils. Whether or not these rocks and soils yield acid wine is another matter again. In other words, in the wine world, the meaning of "acid" depends on the context.

In recent times, concerted international efforts have been made to rationalize and harmonize igneous rock terminology. Rigorous new schemes have been devised in which, as just explained, the terms *acid* and *basic* have been dropped. Modern geologists who are doing research on igneous rocks find it useful to have this new precision and consistency; so far, however, it has not penetrated the wider world. Even geologically inclined readers of this book will have little familiarity with, for example, the rock now known as a *foid monzodiorite*. So, although I adopt the terms *felsic* and *mafic* here, I'll stick with a simple, time-honored classification scheme that leads to familiar rock names, the ones seen in descriptions of vineyards (Figure 4.2).

	FELSIC (ACID)	INTERMEDIATE	MAFIC (BASIC)	ULTRAMAFIC
Main minerals	Quartz Potassium feldspar Sodium-rich plagioclase feldspar	Amphibole Sodium and calcium-rich plagioclase feldspar	Pyroxene Calcium-rich plagioclase feldspar	Olivine Pyroxene
Minor minerals	Muscovite Biotite Amphibole	Pyroxene Biotite	Amphibole Olivine	Calcium-rich plagioclase feldspar
FINE GRAINED	RHYOLITE	ANDESITE	BASALT	PERIDOTITE
COARSE GRAINED	GRANITE	DIORITE	GABBRO	

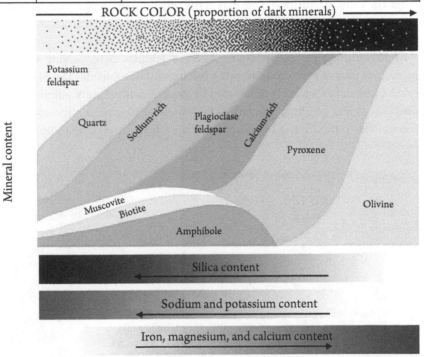

Figure 4.2 The names of the main kinds of igneous rocks. Note that the elements depicted at the bottom of the diagram are locked into the geological minerals and require weathering processes to make them available as potential vine nutrients.

Some Rocks You May Meet in Vineyards

Dominating the felsic side of the scheme is the coarse-grained rock **granite**, which is widespread and well known. Granites are pale in color, with quartz easily visible as clear, glassy, shapeless patches between the chunks of pale feldspar, which may well have cleavage planes catching the light (Figure 4.3a; see Plate 5). (The "black granite," dark blue "granite," and the like, of kitchen countertops and tombstones geologically isn't granite.)

Follow the IP3 up the Dão Valley from near Coimbra, Portugal; drive on Route 79 into the hills east of Temecula, California; strike east on any of the roads leading from the R44 toward the Helderberg, Stellenbosch Berg, or Simonsberg: you will find that the sandy, gravelly soils all around are arrestingly pink because of the rosy potassium feldspar in the underlying granite (Figure 3.5). Pale pink is a common color for granite, as seen also in Beaujolais, France, and the southern Strathbogie Ranges in Victoria, Australia. Pale gray is the other common tint. The whitish feldspars of the Sierra Nevada granite in central California make this rock and its detritus appear pale-gray.

Besides the quartz and feldspar, different granites have varying but minor amounts of minerals such as muscovite, biotite, and hornblende. The Ajaccio wine region of western Corsica is dominated by granites with slightly differing mineral constituents and resulting differences in resistance to erosion. The escarpments around Petreto-Bicchisano, for example, on the road from Ajaccio to Bonifacio, are

Figure 4.3 Examples of igneous rocks found in vineyard areas: (a) granite, near Alcafache, Dão, Portugal; (b) andesite, Tokaj, Hungary; (c) gabbro, Le Pallet, Muscadet, France; (d) basalt, Badacsony, Hungary; (e) porphyry (in a polished slab in the Vatican).

due to such different granites. So, incidentally, are the megalithic monuments that are scattered throughout this region, including in some vineyards. Just to the south, the village of Sartène is built entirely of the local granite.

The massive nature of granite allows it to yield large blocks and slabs which, because of the intergrown texture of the minerals, can take a fine polish, so it makes a celebrated decorative building stone. The famed *lagares* in which grapes for port wine are trodden by foot are traditionally constructed from the local granite. Also, granite is surely one of the more evocative rock names. The word is practically synonymous with ruggedness and durability: it is perhaps no coincidence that many bank buildings are veneered with granite slabs. Arnold Schwarzenegger's jaw "is a block of granite"; Virginia Woolf saw "truth as something of granite-like solidity." However, let's note that the word is commonly used outside geology for any rock that is tough. Many a granite in a mason's yard is geologically not granite, and, it has to be said, the same applies in some vineyard descriptions.

From a distance, **rhyolite** can look like granite, but close up the grain size defines the difference. Occurrences of rhyolite are fewer and smaller scale than those of granite because magmas of this felsic composition are very viscous, and so relatively rarely do they become extruded and rapidly chilled. Nevertheless, viticultural examples of rhyolite do occur—for example, in the Heerkretz vineyard of westernmost Rheinhessen, Germany; in southernmost Arizona, east of Tucson; and at Tokaj, in Hungary. The rocky knoll of Stag's Leap in the Napa Valley consists mainly of rhyolite. In fact, unusually for rhyolite, in the Napa Valley the rock has been much used as a building stone, with a number of wineries being constructed from it. The imposing Greystone building at the north end of St. Helena, which for a time was the Christian Brothers winery and is now the headquarters of the Culinary Institute of America, is made of rhyolite. Just across the valley to the east is a winery located in an old rhyolite quarry.

A variation on the theme of rhyolite is **obsidian**, a fascinating rock, important in archaeology. It tends to be felsic in composition and often comes about through rhyolite lava being quenched in cold sea or lake water. The instantaneous solidification gave insufficient time for any crystal structures to form and so it is a natural glassy (noncrystalline) solid. It is not included in Figure 4.2 because it is fairly rare. However, I mention it here because it does occur at least in the vicinity of some vineyards, in which case promoters invariably make something of it. The name appears on a number of wine labels, from the Mayacama Mountains above the Napa Valley, McLaren Vale in South Australia, and Waiheke Island, off Auckland, New Zealand. Where water became incorporated during the quenching, a less dense, fractured solid results, termed **perlite**. It has the important property of expanding hugely on heating, as the water tries to escape, which results in a fluffy material used to boost soil structure, to filter wine, and to start vine cuttings.

Rocks with a composition falling between felsic and mafic are labeled **intermediate**. The group is dominated by **andesite** (Figure 4.3b; see Plate 5), which is now a well-established rock name but one that had troubled beginnings. In 1835, the German geologist Leopold von Buch wrote a paper about Bolivia and Chile in which he introduced a new word: a volcanic "caldera." In passing, he proposed that the lavas of the caldera should be named after those countries' mountains—hence, andesite. Controversy followed. In essence, he was taken to task about the rocks being insufficiently different to justify a new name, and so the word languished. Not until 1861 was it revived, with a more rigorous justification, by Justus van Roth, after which the name became accepted. (Von Buch, incidentally, was more immediately successful with his word "caldera," as he was with another term he helped establish, a period of geological time which he argued should be called "Jurassic.")

As it turns out, andesite lavas are abundant not only in the Andes but all round the Pacific and in southern Europe as well. On Santorini, the volcanic island in the eastern Mediterranean, the soil is exceedingly stony, some of which is a rubble composed of andesite. The Red Hills AVA, in Lake County, California, takes its name from the striking red soils of weathered volcanic rocks, much of which is andesite, while the Naches Heights AVA of Washington State takes its name from its elevation above surrounding land, on a plateau of andesite lava.

The coarse-grained intermediate rock **diorite** is not widespread, but it makes, for example, the notable hill of Brouilly in Beaujolais. The vineyard soils here are peppered with fragments of this attractive blue-gray rock, larger pieces of which have been gathered to make the vineyard walls and the houses of the workers. It is coarse-grained, which, together with other observations, leads geologists to regard it as an intrusive rock. It is therefore incorrect, as with all *intrusive* igneous rocks, to call the hill a volcano, as some Beaujolais wine books have it. Incidentally, neither is it correct to refer to it as a *prehistoric* volcano. Erosion in recent geological times has given the hill a conical shape, but the igneous activity took place almost 400 million years ago, long, long before "prehistory," a period that by definition involved humans.

The rock **porphyry** (the adjective is **porphyritic**), as we have already seen, has two or more contrasting grain sizes (Figure 4.3e; see Plate 5) and so does not fit readily into the scheme of Figure 4.2. Most porphyritic rocks are intermediate in composition. The name comes from the Latin *porpora* meaning purple, but the rock is found in a wide variety of colors, particularly a deep red and a rich green. It is a most venerable decorative building stone, much revered by the ancient Romans, although the exact location of their main quarry became lost in the mists of time. Victorian enthusiasts repeatedly sought the legendary place, scouring the Arabian deserts until the indefatigable Gardner Wilkinson found some discarded pillars of porphyry in Egypt. He then spied on a distant hillside a patch of purple: "on arriving there . . . my delight knew no bounds, the ground being strewn with pieces of the most sumptuous porphyry while a little farther on was the actual pitched slide down which the blocks came." He had rediscovered the fabled *Mons Porphyrites* or Mountain of Porphyry, the lost site of the Romans' mining operation.

German winemakers esteem soils derived from porphyry, for instance, in the Rheinhessen, Nahe, and Hessische Bergstrasse areas; there is even a Riesling wine named after the rock. The area around Bolzano in the Alto Adige in northeast Italy is carved into a plateau of porphyry (the largest in Europe). The rock is strikingly red in the vineyards to the north, in the Terlan district, while to the south, in the Val di Cembra, the vineyards on the valley sides are interspersed with quarries actively extracting a darker porphyry, for use as a decorative stone.

The mafic part of the classification scheme brings us to **basalt**, perhaps the best known igneous rock after granite. Perhaps you like the word? Well, you could buy a basalt leather jacket or a child's basalt car seat, or maybe a Porsche finished in basalt; one of the most collectible lines of Wedgwood, dating from 1768, is called black basalt. The rock is widespread, very dark gray-green in color, and often almost black. It's hard to make out the individual crystals in this fine-grained rock, but in the sunshine there may be a slight glittering appearance as cleavage surfaces of the plagioclase feldspar and pyroxene catch the light.

The iron from the mafic minerals in basalt gives a characteristic red color when the rocks weather. A chunk of basalt in a vineyard soil may well have a red rind coating its exterior. In fact, soils derived from basalt can sometimes be strikingly red. Potentially, basalt offers a good range of mineral nutrients, but the degree of weathering is important. It has to have progressed sufficiently for the pyroxenes to yield high CEC clay minerals, such as montmorillonite, in order for the nutrients to be available to vines. This is why in active volcanic areas, the soils on the oldest lava flows have the richest vegetation.

The largest volcanic outpourings on Earth are those of basalt, though most are hidden deep below the ocean waves. They occur where tectonic plates are diverging, building ocean ridges and occasionally islands such as Iceland, and around localized hot spots such as Hawaii and the Galapagos Islands. Many of the eruptions are along fissures in the Earth's crust, which spew enormous masses of **flood basalt** in what are sometimes called **shield volcanoes**. Examples on land include the Columbia River basalts of the northwestern United States and the Deccan Plateau of the Maharashtra region in western India. Here, some of the most ambitious winemakers in the growing wine culture of India have established their vineyards up on the so-called Deccan traps, at the higher altitudes given by the 2-kilometer-thick series of basalt lava flows. They erupted around 65 million years ago, interestingly enough around the same time that the dinosaurs became extinct. Similarly huge, the Paraná basalt plateau underlies parts of Paraguay and emerging vineyard areas in southern Brazil and Uruguay.

Basalt lavas and their weathered products are already important in the enormous Columbia Valley AVA, and, although access to water is a constraint in the more arid parts, newer vineyards are increasingly being sited on the lavas. One reason is that recent research here has suggested that the dark basalt can give some thermal advantages to vines. Other examples of basalt associated with vineyards, some with the rock named in the wineries and their range of wines, occur in the Killarney district, King Valley and Macedon Ranges of Victoria, Australia; the Galilee region of Israel (including the Golan Heights); the Côteaux du Languedoc, France; the Pechstein ("tar stone") vineyard at Forst, in Pfalz, Germany; and several parts of Hungary. The last-named instance includes the cone of Somló Hill (Figure 8.4), the tough basalt of which rises abruptly from the flat, sandy plains all around it, the black basalt of Szent György, and the flat-topped Badacsony Hill, overlooking Lake Balaton (Figure 4.3d; see Plate 5).

Gabbro, the coarse-grained equivalent of basalt, is not common at the Earth's surface, although at least a couple of wines are named after it. It is a dense, dark gray to green rock. It can appear almost black, as in the stunning polished wall of the Vietnam Veterans Memorial in Washington, D.C. (although it's officially referred to as granite). The name was introduced into geology by Leopold von Buch, of andesite fame, for some building stones he saw in Florence, Italy, which he thought had come from the little town of Gabbro, about 75 kilometers west of Florence, near the coast (though ironically, the rocks in the hills around Gabbro would now be called serpentinite!). Sometimes found with gabbro is the curious rock at the extreme right of the classification chart (Figure 4.2), the "ultramafic" **peridotite**. It is curious because it presents something of an irony. On the one hand, peridotite is rare at Earth's surface, often just a minor associate of gabbro, and on the other hand, it is volumetrically just about the most abundant rock in the planet as a whole. This abundance comes about because peridotite dominates the composition of at least the outer 400 kilometers of the Earth (it is debatable to what extent the material below that can reasonably be called rock as we know it), though it only rarely emerges through the crust and becomes exposed at surface.

Wineries in a few areas like to mention that these rocks are involved in their vineyards, such as gabbro in the North Yuba part of the Sierra Foothills AVA, and peridotite in the Rogue River AVA in Oregon and at Fraissé des Corbières in the Languedoc. But nowhere more so than in Muscadet, in northwest France. In recent years, some winemakers in this region have begun to trumpet the particular igneous and metamorphic rocks in their vineyards, such as the gabbro that makes up the gently rolling countryside between Le Pallet and Gorges, in the Sèvre-et-Maine wine district (Figure 4.3c; see Plate 5), and the altered peridotites that form parts of the Butte de la Roche, just east of Nantes.

Thrown from Volcanoes: The Volcaniclastic Rocks

Geologists have sophisticated schemes of grouping the ways in which volcanoes erupt and the kinds of structures and materials that result. For our purposes, however, we can simply distinguish between a relatively calm, oozing out of lava, with little material being hurled into the air, and an eruption in which clouds of material are ejected, sometimes explosively. It all hinges on the silica content of the magma because of its effect on lava viscosity. Where the silica content is low, the lavas tend to be fluid and to produce runny basalt flows, as in basalt shield volcanoes. A magma with an intermediate silica content again can produce lavas, as we have just seen for andesite, but these are more cindery. Gas pressure can build up in them, leading to explosive activity and to disintegration of the material into solid fragments. Eruption of rhyolite material will be dominated by material that is catastrophically fragmented and hurled into the air because the silica-rich magma is so highly viscous.

Andesite and rhyolite volcanism is characteristic of convergent plate boundaries, where long mountain belts like the Andes form and there are chains of oceanic islands like the East Indies and Japan. The violent behavior of Mount St. Helens, Pinatubo, Vesuvius, and Krakatoa is legendary. These are generalizations, of course: we all know of the ash clouds sometimes emitted by Icelandic volcanoes, even though they are typical shield volcanoes with basalt lava flows. With an andesite or rhyolite volcano, much of the fragmental material and lava erupts from a central throat and doesn't travel far. And so we get the popular image of a volcano, conical in shape around a central vent. To contrast with basaltic shield volcanoes, they're called **cone volcanoes**.

We now look more closely at this fragmental material. The subject straddles this and the next chapter because, although it is igneous in the sense that it solidified from molten magma, it settles from the air onto the Earth's surface exactly in the manner of a sediment. We will deal with these fragmentary volcanic products here and call them **volcaniclastic** (alternatively, **pyroclastic**) materials. There is a profusion of names here, reflecting the great interest in volcanoes and the range of materials they produce, but we will look at just a few products, those met most commonly in the wine world.

First, all these volcaniclastic materials are sometimes grouped together under the single word **tephra**. A fine example of its use is in summarizing the wide range of fragmental material that has a long history of eruption in Campania, Italy, in the volcanic complex more or less centered on Naples. The region has, of course, long been associated with viticulture. For the poet Martial (c. 40–c. 104 A.D.) it was "verdant with shading vines" and "the noble grape loaded the dripping vats"; Pliny the Elder's "*Natural History*" tome (A.D. 77) records that the region's wine "whether by means of careful cultivation or by accident, has lately excited consideration". Just two years later, though, everything changed with the violent eruption of Vesuvius. Martial watched the horrific events and recorded that now "all lies drowned in fire and melancholy ash", with "piles of ashes, spreading all around", "undistinguished heaps deforming the ground".

In Naples itself, the conspicuous hill that is capped by the San Martino monastery dominates the city and since 1700 has had a vineyard clinging to its vertiginous slopes. Its striking terraces are carved into layers of tephra of different kinds. The oldest is the gray ash that forms the base of the hill. It was deposited 40,000 years ago (during an eruption that may have contributed to the demise of our cousins the Neanderthals); the youngest erupted 12,000 years ago and is a fine yellow ash at the very top of the hill, just below the fortress of St. Elmo.

Just to the west are numerous further volcanic centers, almost all with associated vineyards. They include Averno, Astroni, and Pozzuoli, and the islands of Ischia and Procida, all built from assemblages of all kinds of volcanic debris, deposited over long periods. Some of the light and dark gray ash layers on Ischia settled over 150,000 years ago—and the region is still volcanically active today. This long, violent volcanic history is aptly summed up by the name given to the whole region, Campi Flegrei—the fields that burn—and tephra is a useful word to cover this wide assortment of ejected material.

Tephra is subclassified according to grain size, just like the fragmental sediments discussed in Chapter 5. A simple scheme is shown in Figure 4.4. We call coarse fragments of volcanic material **bombs** if they show external signs of having been fluid, such as a twisted shape or corrugated surface. **Blocks** are more angular and, in line with the terminology for sedimentary rocks, make the rock called **volcanic breccia**. Coarse material like this does not travel far when it is ejected, leading to accumulations around the volcanic vent and over time building the characteristic shape of a cone volcano. Volcanic breccias occur in areas of violent eruptions, the Campi Flegrei, for example, or the island of Santorini, and are striking in parts of the northern coast of the Island of Madeira. There, blocks of dark basalt over a meter across occur in layers visible in the crags that overlook some of the vineyards, and are conspicuous in the sea cliffs. The Chateaugay vineyards in the Côtes d'Auvergne are largely founded on volcanic materials, which are also much trumpeted as the source of Volvic bottled water. In just one rock face, near the Volvic railway station, a whole variety of volcanic chunks is visible, including some with a bulbous shape

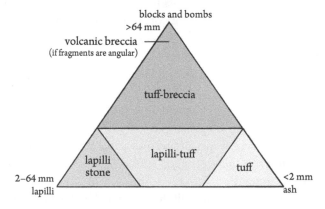

Figure 4.4 Some names of different kinds of tephra, the material thrown from volcanoes, showing how they vary with particle size.

that reflects their fluid origin. French workers have termed these bulging volcanic bombs *chou-fleurs*: cauliflowers!

Lapilli are roundish granules that in many cases were ejected from a volcano as semi-molten lava, cooling as they fell back to Earth. The accumulations of tephra mentioned in the examples above include layers of lapilli, but there's a particular use of the material in viticulture on Lanzarote, in the Canary Islands. Lanzarote is remarkably windswept and dry, with one-eighth the rainfall of Bordeaux, and in an especially dry year, little more than twice that of the Sahara. Consequently, the basalt weathers slowly, but the islanders have evolved a highly labor-intensive but pragmatic way of improving vine nutrition and hydration. They scoop hollows down into the tephra to reach the older, more weathered material below, and then they use blocks of lava to build semicircular low walls on the windward sides of the scoops. A vine planted in the hollow thus receives some wind protection, and manure placed around its roots, replaced every couple of years or so, provides essential nitrogen and phosphorus. The clever feature, though, is that a mulch of lapilli is then applied, no more than several centimeters thick, the optimum for conserving precious moisture. How could such a specific procedure have been discovered?

In 1730, Lanzarote's volcano began to erupt and for over six years proceeded to blanket the island with lava and tephra. The islanders had to be evacuated, and from nearby could only watch helplessly as their villages and fields were gradually destroyed. But later, when an emissary was sent from Madrid to assess the damage, he noticed a curious thing. The lava and thick layers of tephra had indeed created *"malpais"* (bad land) but in areas with only thin deposits of lapilli, vegetation was thriving. So when people eventually began returning to Lanzarote they utilized this observation, by planting crops and spreading a thin—no more than a few centimetres—layer of lapilli on them. It worked. Astonishingly, the crops flourished, harvests multiplied, and before long the island's population was burgeoning. And for the first time ever, the island even began producing wine!

Research has shown that lapilli provide the best material for such a mulch, giving a good surface area to attract dewfall at night, and a balance between high permeability, which allows the water to trickle through to the soil below, and a sufficiently tight packing of the grains to minimize its evaporation out again. Some water may be stored in the underlying tephra, which can have porosities as high as 33%, but the nub of this system is the mulch. Visitors remark on the verdant vines that are apparently thriving in a moon-like expanse of black grit, (though, as with the rocky vineyards elsewhere, the root activity is taking place in the material hidden below). Nowadays, the rock mulching that was discovered on Lanzarote is a world-wide industry, and is utilized not only in vineyards and other forms of agriculture around the world but in domestic gardens as well and, who knows, perhaps in that flower pot on your window sill.

The finer categories of volcaniclastic grain size are fairly self-explanatory. Any area with recent volcaniclastic deposits is likely to contain **ash**, the very finest

material. This includes all the examples mentioned earlier, as well as places such
as Crete, Etna, the plains stretching way west of Melbourne, Victoria, and the eastern part of the Soave hills, east of Verona in northeast Italy. On becoming rock, by
the same processes as explained for sedimentary rocks in Chapter 5, it is known as
tuff. Sometimes we add the adjectives felsic and mafic, according to their pale or
dark color, respectively, and by inference their silica content. Tuff is found across
Campania, Italy, but a noteworthy example, known now in its Italian form as *tufo*,
occurs by the river Sabato about 10 kilometres north of Avellino. Here, in about the
10th century, the underground workings of this quality building stone and associated sulfur deposits were greatly enlarged, consequently the nearby town grew and
eventually was to take its name from this desirable rock: Tufo. In turn the vines that
had grown there for centuries absorbed the name, and so we now have the grape
Greco di Tufo. It's an unusual instance of the contribution of geology to wine!

In several areas of northern Hungary, extensive underground cellars have been
carved into the relatively soft tuff that abounds there (Figures 4.5 and 4.6). The
town of Miskolc alone has nearly 600 of them. In Eger, home of the famous Bull's
Blood red wine, the rows of cellars to the west of the town with their creatively deco-
rated entrances are a major feature for tourists. The wine tastings they offer enhance
the attraction, of course, and the name of the street, Szépasszony-völgy, or Valley of

Figure 4.5 Felsic tuff, near Eger, Hungary. Larger fragments of plagioclase feldspar and
some pumice give a porphyritic texture to the otherwise fine-grained rock.

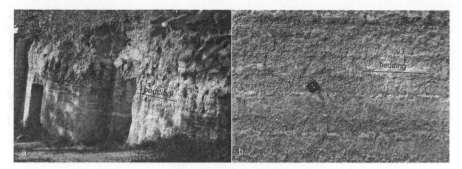

Figure 4.6 (a) Wine-storage caves carved in soft, bedded tuff, Kácz, northeastern Hungary. (b) Bedding (or stratification; see Chapter 5) is indicated.

Beautiful Women, probably helps too. However, in much wine literature, the rock here is mistakenly referred to as *tufa*, which is geologically quite different from tuff and has nothing to do with volcanoes, a confusion I discuss further in Chapter 5.

Pumice is a fine-grained, typically pale-colored rock, distinctively full of bubbles which give it a foamy look and a remarkably low density—such a low density that it floats! It has nothing to do with pomace, the residue left from pressing grapes (though in the United States pomace seems often to be called pumice). The rock is produced when extremely hot and pressurized gas-rich lava is rapidly ejected from a volcano, such that in the reduced pressure at surface, the gas forms bubbles, and the mass becomes quenched, often by landing in water. It's solidified lava froth!

Although large blocks of the material are common in the vineyards of the classic pumice location of Lipari, the Aeolian island just north of Sicily, in most vineyard soils it exists as fragments of sand or lapilli size. Such pumice fragments are common in the Waikato and Bay of Plenty regions of North Island, New Zealand, and near Mád, to the northwest of the town of Tokaj, Hungary. On Tenerife in the Canary Islands, pumice is used as a mulch in the same way as the lapilli are on Lanzarote. Around the vineyards in the hills north of Sebastopol, Sonoma County, California, there are layers rich in pale pumice fragments, as there are to the east, along the road that climbs from Santa Rosa over the Mayacamas Mountains. The peculiar properties of pumice make it useful in the construction industry, where light weight is needed, and as a mild abrasive. The roof of the Pantheon in Rome involves pumice in the concrete and is still, two thousand years after it was built, the world's largest unreinforced concrete dome. And that ineffable look of your stone-washed jeans is due to pumice.

The range of minerals in most of these porous, volcaniclastic materials means they can support a diverse microbiota and yield a good range of mineral nutrients. Recent research on Gran Canaria has revealed that the island's tephra contain microscopic holes (former gas bubbles) now full of fresh water in which diatoms and other microbes flourish, including the bacteria that help make nutrients available to the vine roots. Any allophane (Chapter 3) weathers particularly quickly to provide nutrients. Thus, some volcanic areas are famous for their luxuriant vegetation; in the Naples region, the Romans talked of the *Campania Felix*—the Fortunate Country. Pliny the Elder wrote in his "*Natural History*" of this region's grapevines scrambling over poplar trees, "embracing their brides and climbing with wanton arms in a series of knots among their branches, rising level with their tops, soaring aloft to such a height that a hired vintager expressly requires in his contract the cost of a funeral and a grave". Remarkable, though together with three harvests a year this probably didn't lead to fine wine.

Finally, it's fashionable to claim that wines from volcanic soils have some sort of commonality, notwithstanding the enormous variation in the chemical and physical properties of volcanic rocks. Some claim they have a distinctive character, such as an unusual complexity and exotic quality. They are even ascribed a peppery, spicy, smoky or—yes—a fiery flavor. So it's worth recalling that volcanic soils are formed from the same kinds of chemical elements and the same kinds of geological minerals as other rocks, and are identical to those intrusive igneous materials that didn't happen to make it to the Earth's surface. There is no special ingredient just because at some past time the soils' parent rocks emerged, either molten or solid, from a volcano.

Further Reading

Douglas Jerram, *Introducing Volcanology: A Guide to Hot Rocks*. Tampa, FL: Dunedin Press, 2011. This work is about volcanoes rather than the resulting rocks, but it's clear, enthusiastic, and well illustrated.

The same author (with Nick Petford) gives more emphasis to the rocks, though with students of geology in mind rather than amateurs, in *The Field Description of Igneous Rocks* (*Geological Field Guide*, 2nd ed.). Hoboken, NJ: Wiley-Blackwell, 2011.

5

Sediments and Sedimentary Rocks

We are on more familiar ground in this chapter, looking at processes and materials found in the world all around us. Even the names of sedimentary rocks are well known—sandstone, shale, limestone, and so on. Clearly, these materials are highly relevant to vineyard geology because more than three-quarters of the land surface is sedimentary in origin: most of the world's vineyard areas are underlain by sedimentary rocks.

Sediment is the detritus produced from the weathering of already existing rocks. (I explore the process in Chapter 9.) Usually, wind, ice, or water soon moves the debris away, eventually to be deposited and then buried beneath further sediment and with time hardened into sedimentary rock. Weathering can also dissolve material, later to be precipitated. And, needless to say, all the sediment in question here is of geological origin; it has nothing to do with the organic sediment that is thrown, say, in a bottle of vintage port!

The Detritus We Call Sediment

Wind and flowing water may be able to pick up sediment and move it, depending on the size of the fragments. Faster-moving currents can carry bigger particles: it's to do with energy, as discussed in the context of rivers in Chapter 8 (see Figure 8.8). The result is sediment **sorting**. We can easily see the results on a beach—a sandy spot here, a pebbly patch there—because the tides and shore currents have moved the sediment around and sorted it. Thus, most detrital sediments have a characteristic grain size, and we use this to classify the material.

The terms for the different sizes are pretty much in line with everyday language: sand, silt, clay, and so on (Figure 5.1). **Clay** is the finest sediment. It's composed mainly of the tiny clay minerals that we met in Chapter 3 and has the smooth, slippery feel and handling properties we're all familiar with; the individual constituent particles are far too fine to see, even with a powerful hand lens. Imagine: if we scaled up a grain of sand to the size of a wine cask, then an individual clay flake

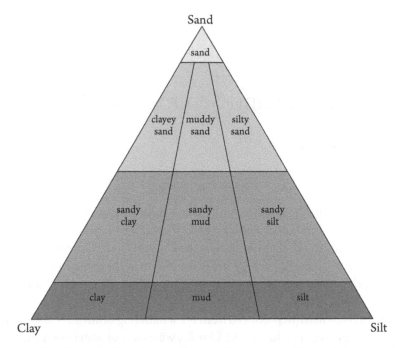

Figure 5.1 Classification of sand, silt, and clay sediments, according to their proportions. Each corner of the diagram represents 100% of the labeled component, and thus the opposite face of the triangle represents 0%. The central fields show mixtures in various proportions.

would be smaller than a coin. A useful word here is the adjective **argillaceous**, meaning clay bearing. Thus, we talk of argillaceous sediments, or argillaceous rocks.

Less fine grained and giving a slight flouriness between the fingers is **silt**. Here, the individual fragments are just visible with a hand lens. The settling out of silt from water is well known, hence the phrase "silting up," which is sometimes a problem in vineyard irrigation systems. But these clayey and silty sediments can be transported huge distances, and much of it will make it out to the deep sea. The technical name for this fine mixture? **Mud.**

Also important for some vineyard regions is silt that was deposited from moving air, the material called **loess**. It covers vast areas of China and the U.S. Midwest. In fact, over 10% of the Earth's land surface is covered by this wind-blown "dust," including some of the most fertile and agriculturally productive regions in the world. The name originates from a German word for the deposits along the Rhine near Heidelberg, where it floors many vineyards today. Loess is readily warmed and is both porous and permeable, though there is usually some clay content that provides a degree of water retention. This also gives loess a degree of strength, soft enough to be easily penetrated by roots but strong enough to hold its shape. Cellars have been carved into the thick loess deposits of the Kremstal and Kamptal districts of Austria

Figure 5.2 Loess at Langenlois in the Kamptal region of Austria. The featureless deposit is sufficiently strong to hold vertical faces cut into it, yet soft enough to easily be carved into storage cellars, here along the Grosser Buriweg.

(Figure 5.2), and it's common in Oregon and Washington State as well as in Tokaj, Hungary. In the thick loess deposits in parts of the Kaiserstuhl, in Baden, Germany, wide terraces have been cut in order to allow mechanization of the vineyards.

The next coarsest fragmental material is familiar to everybody: **sand**. But let's note that it's the particle size that defines it, not its color, which usually depends largely on the local bedrock. Northern Europeans may think of sand as buff-yellow in color, but a Caribbean islander thinks of it as white. Sand on Hawaii tends to be black because of all the lava, though there is a beach on the island that is green because of the mineral olivine that is concentrated there. The Ile de Groix off the southern coast of Brittany, France, has some beaches with a distinct blue cast because of the accumulation of a deep blue amphibole mineral (called glaucophane) that is weathering from the metamorphic rocks of the island.

In everyday parlance, a number of words—including grit, granules, stones, and shingle—are used for fragments that exceed sand grains in size. Geologists call all this material **gravel** (Figure 5.3; see Plate 6) and delimit pebbles, cobbles, and boulders according to size. A single kind of mineral, say a fragment from a vein of quartz (Chapter 7), could form a pebble and maybe even a cobble, but the vast majority of fragments of this large size will be composed of some preexisting rock. They are

Figure 5.3 Examples of pebbly vineyard soils. (a) Beychevelle, Saint Julien, Médoc, France; (b) Newhouse, Snipes Mountain AVA, Washington.

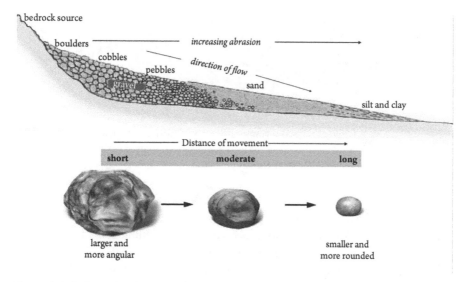

Figure 5.4 Relationship between the size and shape of fragmental sediments and the distance of movement.

usually well rounded; in general, fragments that have moved more have had more chance to bash into each other and knock any rough corners off, through a process called **abrasion** (Figure 5.4). All these ideas are shown in Figure 5.5, a reproduction of a pocket guide that geologists use for recording the characteristics of a sediment.

The pebbles in a gravel provide excellent drainage but are themselves largely inert. That's why they have survived. Commentators like to mention what the pebbles of, say, various vineyards in the Médoc are made of, but while this may be of academic interest, it's irrelevant to the vines: what matters are the variations in fragment size and packing that affect the drainage properties of the soil. Writers in English sometime use the French word **galets** for water-rolled fragments, from the iconic quartzite

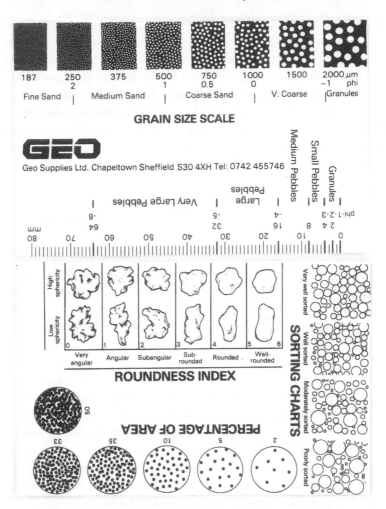

Figure 5.5 Reproduction of the two sides of the kind of pocket-sized card that geologists use when recording characteristics such as grain size and shape of a fragmental sediment. This example was kindly provided by Geo Supplies of Sheffield, England.

pebbles and boulders of the southern Rhône, especially at Chateauneuf-du-Pape. In parts of the Coteaux de Languedoc, the galets are formed from granite and in Boutenac, Corbières, from a brown-stained quartzite; Walla Walla, Washington, has old river channels filled with galets of dark basalt. Incidentally, the pronunciation of galet is roughly GALay. You don't stress the second syllable, nor do you pronounce the "t" (unlike in roulette) unless you want to say the soil is made of cake.

In general, moving ice doesn't sort the debris it carries. Ice can look dirty because of the finely milled grains and clay that are dispersed in it. Yet it may simultaneously be carrying particles of any shape and size, including enormous boulders.

On melting, it's all just dumped, in an unstratified, chaotic jumble. Rumor has it that a professor on a field trip was pointing out examples of huge boulders left behind after the melting ice had retreated when a student asked where the glaciers are now. "Well," said the professor, "they've gone back to get more rocks." There are ways other than melting ice in which such a hodgepodge of a deposit can arise, for instance, through submarine landslides and volcanism, but where it is definite that the deposit is due to ice, we call it **till**. Such sediment is important in vineyard regions that were recently glaciated geologically, such as Long Island and the Finger Lakes of New York State.

Easily Overlooked: The Dissolved Component

We readily see weathering processes breaking down rocks into fragmental detritus, but we tend to forget that at the same time material is being dissolved. For example, because rainwater is usually slightly acid, it reacts with carbonate rocks. When the water moves away, anything dissolved in it is invisible; we see it only if conditions change such that it is precipitated out of solution. And if that happens, we have a **chemical sediment**.

Precipitation can occur on the floor of an ocean or lake, in a cave, or where a spring flows out onto Earth's surface. Where shallow seas occur in arid areas and the water outflow becomes restricted, such as in the Red Sea and Persian Gulf today, evaporation leads to an increasingly saturated brine, and minerals begin to precipitate out, typically calcite, gypsum, halite, and potassium salts, in that order. We call such deposits **evaporites**. They also form in restricted inland seas, though the adjacent land tends to be highly saline and hence inimical to vines. Even so, there are indications dating back millennia of vines growing near the Caspian Sea, and there are successful vineyards today near the Dead Sea, in Jordan and Israel.

Of more widespread viticultural relevance are the results of a seafloor acquiring a thin veneer of precipitated calcium carbonate. Seawater has significant concentrations of dissolved calcium and carbonate ions, so changes in physical conditions can trigger their precipitation, producing a **calcareous mud**. When this mud hardens, it helps form the common vineyard rocks limestone and marl. A classification of calcareous rocks is shown in Figure 5.6.

A Contribution from Biology

The oceans contain vast numbers of microscopic organisms. Some are made of silica, such as **diatoms**—tiny cells of algae said to be floating in the oceans in such numbers they account for a quarter of all the photosynthesis on Earth—and the beautiful glass-like architectural masterpieces called **radiolaria**. Other organisms,

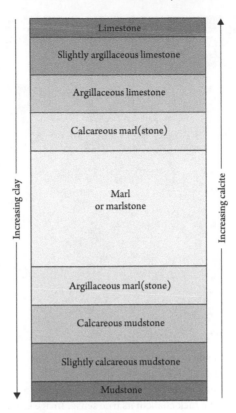

Figure 5.6 Classification of calcareous rocks.

such as **foraminifera**, are made of calcium carbonate. Their broken remains, along with things we readily see such as corals and sea shells, are major contributors to calcareous sediment. Particularly noteworthy are the calcium carbonate scales of plankton called **coccoliths**, not only because blooms of them in the ocean can be huge—visible from space—but because they make that celebrated rock: chalk (Figure 5.7). Sometimes, tiny fragments of all these creatures may roll around on the seafloor and become successively coated with chemical precipitates of calcite, giving the tiny egg-shaped bodies which, from the classical Greek word for egg, are called **ooids** or **ooliths**. The sediment (and the rock formed from it) is an **oolite**.

All geology students learn about the Bahamas region, because the shallow, sun-lit, moderately warm seas typify the conditions where many calcareous sediments form. Those gorgeous white Bahaman beaches are largely made of calcite—each grain a fragment of a coral, a shell, a sea-urchin spine, or a glob of microorganisms. Most of the material is sand-size, pale gray-pink in color, but there can be muddy patches and coarser patches, maybe even areas of pebbles.

Just offshore, the waves do a great job of sorting the grain sizes. Further out, any-one who has snorkelled in such areas will have been able to see (beyond the fish!)

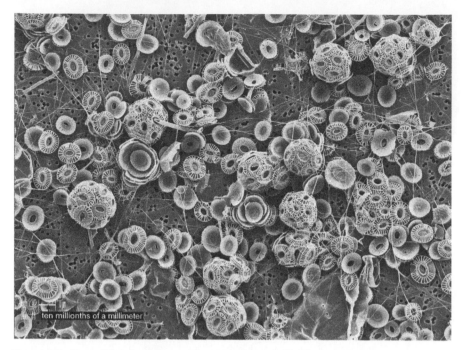

Figure 5.7 Coccoliths, the main component of true chalk. The round objects are a plankton characterized by its enclosing curved plates or scales, many detached examples of which are scattered around the picture. The plates are called coccoliths, and they are the feature that characterizes the rock chalk in its true sense. Image kindly provided by Dr. Alex Poulton.

just how variable the seafloor is. Imagine: as we swim there may be smooth patches of gray mud below us but with a slight depression that has allowed silt to accumulate, and over there currents are making ripple forms in sand. Beyond, there are some level patches of oolitic sediment, and beyond them the disintegrated remains of an old reef system, with localized mounds of biological debris. In other words, although from a distance the sea bottom may look homogeneous, in detail it's varied and constantly changing. Thus, although sedimentary strata along a hillslope may look uniform, as soon as we zoom in, we see all sorts of local differences. It may all sound rather academic, but it's the key to understanding the localized, fine variations in the bedrock and soils of vineyards around the world that are underlain by sedimentary materials.

Hardening Sediment into Rock

Where sediment is accumulating, the lowermost, first deposited sediment becomes increasingly buried and will undergo a series of processes that will harden it into rock. We call all these changes **lithification** (Figure 5.8), and the sediments are said to become **lithified**. The word comes from the classical Greek word for rock, *lithos*, the same root we use in terms like lithography and monolith.

A number of processes work hand in hand to lithify sediments. First, there is **compaction**. I explain this process in Chapter 10 with reference to soils, but it's also important with reference to argillaceous sediments. Here, with all the time available in nature, electrostatic bondings start to develop between the increasingly touching particles, such that eventually the material becomes hard enough to be called a rock.

Silts and sands show relatively little compaction but another process may be starting: **cementation**. The pores of most sediments will contain water which will

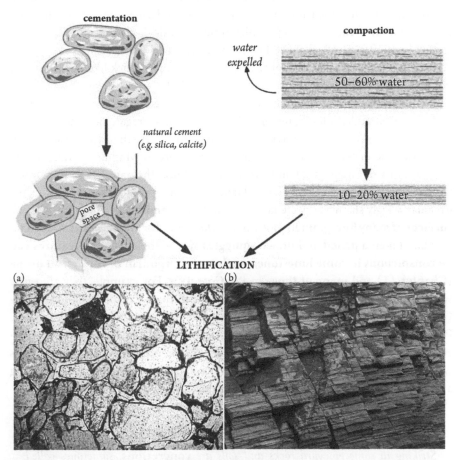

Figure 5.8 Lithification, or hardening sediment into rock, consists of some combination of cementation and compaction. Cementation dominates in sediments with coarser fragments, where a natural cement precipitated from percolating waters bonds grains together (as seen here in (a), a microscope view); compaction dominates in finer sediments, such as producing a mudstone, shown here in (b).

have dissolved constituents in it, so with the slightly changing conditions at different depths in the sediment pile, some of it might just become no longer soluble. A small amount of some mineral precipitating out will attach itself to part of a sediment grain and perhaps coalesce with the material starting to film an adjacent grain such that the two particles become cemented together. As the process proceeds and pervades the sediment, the grains increasingly become locked. It's not a matter of filling in *all* the pores. In fact, rarely does this happen, instead just providing localized bonding between grains while leaving most of the pore space intact. This is why hard, well-cemented sandstones can still have lots of pores and be good aquifers (Chapter 10).

The solubility of calcite (relative to silica) enables it to dissolve and precipitate such that calcareous sediments can lithify relatively rapidly through cementation. It can even happen at the land surface: limestones bordering some Florida beaches contain within them Coca-Cola bottles from the mid-20th century. Such near-surface precipitation of calcite is the origin of cave formations such as stalagmites and stalactites and also, as a purely surficial coating, of the rock-like appearance of objects caught up in the carbonate waters of so-called petrifying wells.

Another curiosity involves the tendency of calcite to dissolve; its solubility increases even under a little pressure. Thus, where an aggregate of calcite grains is being buried, the downward burial force at the points where the particles are touching causes the material to preferentially dissolve there, leaving impurities such as black iron oxide (magnetite) and carbon as residues. They are left as thin seams which, because the process operated differently from grain to grain, have a highly irregular, wiggly shape. They are known as **stylolites and are visible on the clean surfaces of many fine-grained limestones. They look for all the world as if someone had taken a pencil and drawn squiggles across the rock. The features can be conspicuous in some limestones, such as the Urgonian of southern Europe (Chapter 11) and many of those in the** Côte d'Or. For example, they are easily seen in the quarries along the D974 road between Aloxe-Corton and Nuits-Saint-George, where the so-called Comblanchien marble (Chapter 6) is being extracted. Particularly striking examples of stylolites occur in the rose-pink limestone from eastern Tennessee. It's in the Lincoln and Jefferson Memorials in Washington, D.C., is widely used for flooring, and is seen in the walls of many a toilet stall, thus providing a good opportunity for us to meditate on the finer points of lithification processes.

Concretions in Sedimentary Rocks

Striking in some vineyard rocks and soils are **concretions,** *sometimes called* **nodules.** *They are distinct roundish lumps, different from and markedly harder than the host sedimentary rock. The bodies vary in size from millimeters to several meters across, most typically tens of centimeters; Figure 5.9 illustrates the range in appearance. Concretions baffled early geological collectors, who thought they might*

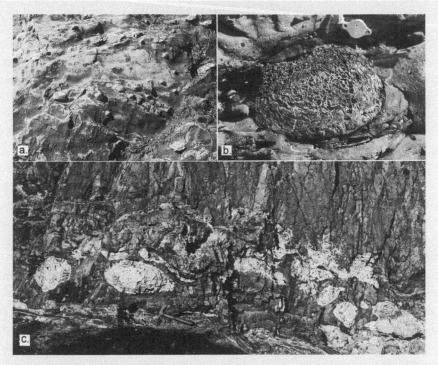

Figure 5.9 Examples of concretions (or nodules): (a) in loess, (b) in mudstone, and (c) chert concretions in limestone.

be such things as ancient artillery, cosmic debris, dinosaur eggs, or even gifts from the divine. Rather more prosaically, today we know they are an effect of early lithification. Essentially, the bodies are due to mineral replacement and intense chemical precipitation within the sediment but localized around some nucleus, typically a mineral grain or a fragment of some organism.

The relative hardness of concretions reflects their having had a head start, so to speak, as the host sediment lithified. Today they commonly weather out as intact loose pellets or balls. Besides flint, concretions commonly involve cements of iron minerals such as siderite and goethite, and carbonates such as calcite. That the lithification processes begin soon after a sediment starts to be buried is demonstrated by the fact that concretions also form in soils, including in vineyards (Chapter 9). More dramatically, there are examples in the coastal marshes of eastern England of concretions formed around nuclei of shells and shrapnel from World War II!

Examples of concretions can be found in the loess deposits of the Willamette Valley, Oregon; in the bedrock of Cahors and Saumur, and in Picpoul de Pinet in the Languedoc in France; in Ribera del Duero and Penedès in Spain; and in one of the oldest wineries in Uruguay, Viña Varela Zarranz. In Italy there are examples around Piacenza, and a wine from the eastern Marche is named after their local name, cacinello.

Rocks from the Detritus

There is no formal terminology for rocks that contain very coarse detritus—boulders and the like. We can talk of a boulder bed, which is nicely descriptive without implying anything about how it formed. If we know the boulder bed is of glacial origin, as discussed above for till, then we can call the lithified equivalent a **tillite**. Most tills produced during the most recent ice age are still unlithified. However, there are fine examples around the world of tillites produced in ancient ice ages, hundreds of millions and more years ago.

Up above the Constantia vineyards of Cape Town, on Table Mountain, is a tillite formed about 450 million years ago. It is also seen to the north at Pakhuis Pass in the Cederberg Mountains, not far from the winery there. And astonishingly, just further up the road from there, near Nieuwoudtville and in the Olifants River wine area, is a completely different tillite, this one having formed during a different ice age, about 300 million years ago. In both cases there is good, independent evidence that these boulder-bearing rocks were associated with ice. Hard though it is to imagine ice sheets in what are now such warm areas, both of these examples formed at times when the continent of Africa was much closer to the Earth's South Pole. However, there are many instances where geologists quarrel about whether or not a poorly sorted, very coarse-grained deposit actually involved ice, so here the term *boulder bed* will do.

Sedimentary rocks that comprise fragments of pebble and cobble size are called **conglomerates**, if the fragments are reasonably smooth and rounded. Typically, the fragments are cemented together by silica, making a tough rock. A picturesque vernacular English name for these rocks, long predating conglomerate, is *puddingstone*, supposedly because the pebbles resemble the plums in an old-fashioned Christmas pudding. And curiously, in the eighteenth century, this very English word sneaked across the English Channel and morphed into *poudingue*, which is now the French for conglomerate. It even got to New England: Oliver Wendell Holmes wrote about puddingstone, and it's now the state rock of Massachusetts! Conglomerates can be very attractive if the fragments are brightly colored. The jagged peaks of Montserrat in northern Spain, for example, conspicuous from Barcelona, are carved in a striking pink-gray conglomerate. In Sonoma County, California, uplands of the resistant Dry Creek conglomerate separate the Anderson and Dry Creek Valley AVAs, and weather to form a coarse, gravelly soil prized for zinfandel vines.

Where the fragments are angular, the rock is termed a **breccia**. The word is Italian, and so the "ci" is pronounced in the Italian way, as the ch in chin. Breccia has been much used in Italy as a polished decorative stone, though etymologists say the word originated in an old High German word for breaking or breaching. The rock is said to underlie a number of vineyards in Italy and elsewhere, but the word appears on some wine labels even where the vineyard is not located on breccia.

For medium- and fine-grained fragmental rocks, the terminology is simple, deriving directly from the sediment names: sandstone, siltstone, and mudstone. Strictly, a mudstone contains a small amount of silt, unlike the finest-grained rock of all, consisting wholly of clay minerals, a **claystone**. All these rocks are commonly found in the world's vineyards, and they would seem to require little further explanation here, with two exceptions.

First, we have that somewhat enigmatic rock **graywacke** (pronounced GRAY-wacky and mostly spelled greywacke outside the United States). The name has always courted controversy: in his pioneering book (1839) on rocks of Silurian age, Sir Roderick Murchison observed that "it has already been amply shown that this word should cease to be used in geological nomenclature." Nevertheless it persists, for example, appearing on wine labels from California's Russian River Valley, South Africa's Franschhoek, and Australia's Barossa. The rock is essentially a very impure marine sandstone, with sand-sized or coarser fragments set in much finer, clay-rich material (Figure 5.10). The argillaceous matter usually gives the rock an overall dark gray look, and the fragments typically consist of angular grains of quartz, feldspar, and perhaps small rock fragments. But how can these two different sizes of particles

Figure 5.10 Graywacke, seen under a microscope, showing the characteristic lack of particle sorting. Angular fragments half a millimeter across of quartz (appearing in black, white, and various shades of gray) and of rocks are interspersed with finer grains and very fine, indiscernible material.

come together? How did they avoid being sorted apart by currents? Such questions puzzled geologists for many years, and it became known as the "graywacke problem."

The story of the 1929 earthquake at the Grand Banks of Newfoundland is famous in the annals of geology: its repercussions shed light on an important oceanic phenomenon, and it helped solve the graywacke problem. The tremor caused considerable damage and loss of life and, puzzlingly, cut off all transatlantic communications with Newfoundland's east coast. As it was later discovered, this was because a long way offshore, far from the site of the earthquake, no fewer than twelve of the submarine telegraph cables had snapped. Why?

It took over twenty years for geologists to collect evidence and piece together the answer. The earthquake waves had shaken and jostled the coarse sediments just offshore and had stirred them up into a turbulent suspension. The resulting cloud of sediment was denser than the water around it and so moved underwater down the slope, gathering finer sediment as it travelled, and gaining sufficient momentum and energy to break the telegraph cables (in twenty-eight different places!). Such a dense, turbulent, gravity-driven mass has become known as a **turbidity current**. Eventually, such currents come to a rest, and their load of shallow-water sand mixed with deeper-water mud settles on the deep seafloor. Subsequent examination of cores drilled into this submarine sediment on the Atlantic floor off Newfoundland showed that it consisted of as yet unlithified graywacke.

The process is not unlike our stamping on the edges of a large puddle in a muddy vineyard and seeing clouds of sediment billow down toward the middle of the pool. Over geological time, such events will be far from unique, and in parts of an ocean prone to earthquakes, thick sequences of these poorly sorted sediments can be built up. In short, graywackes are the result of turbidity currents.

The rock name came originally from the Hartz Mountains of Germany, as *grauwacke* (gray sandstone), but it was soon introduced into English, as greywacke, though with various nuances of meaning. Partly because of this imprecision, and because modern classification schemes prefer the label "poorly sorted argillaceous sandstone," the term continues to slip into disuse among English-speaking geologists. But there is a conspicuous exception: New Zealanders love the name! Geologists there use it to cover both the range of hard gray rocks that dominate the central bedrock spine of both the North and South Island and much of the gravelly river detritus in areas such as Hawkes Bay, Marlborough, and Wairapara. Kiwi children are weaned on the word: in a series of popular tales, there's a cunning cat called Greywacke Jones. Rugby teams compete for the Greywacke Trophy. It has even been suggested that graywacke should be New Zealand's national rock! It's hardly surprising, then, that the rock lends its name to one of the country's most prominent wineries and range of wines.

The second exception requiring further explanation is **shale**. Down at the fine end of the fragment-size scale we have mudstone, but there is a variant on it, the rock shale. This material comprises fully 46% of the sedimentary rock at the Earth's land surface and hence, presumably, is widespread in vineyards, though wine writers seem to make little of it. Shale is soft and it is **fissile**. That is, it tends to split, though in a very irregular way. When argillaceous sediments are settling, slight compositional variations may produce microscopic-scale layers, and the flaky shape of clay minerals makes them settle parallel to the seafloor. It's like throwing a plate into water; it may start by sinking vertically, but it is more than likely to end up flat. The resulting tendency to splinter in a direction that is overall parallel to bedding, the **fissility**, is the characteristic of shale. It distinguishes shale from mudstone, which has no tendency to split, and from the metamorphic rock slate, where the splitting is more organized, more regular, a cleaner break, and can be at any orientation within the rock.

Shale is a weak, crumbly rock. Its softness was noted as long ago as 1683: Martin Lister noted in his "ingenious proposal for a new kind of map" (the dawning of the idea of a geologic map) that shale is "soft like butter to the teeth and with little or no grittishness." Because it's defined by its physical characteristics, it can be of any chemical/mineral composition. Some shales around the Finger Lakes of New York and the west side of Paso Robles, California, are calcareous in composition; some in the Heiligenstein, of Kamptal, Austria, and the Mayacamas Mountains between the Napa and Sonoma valleys originated from volcanic parents and are rich in iron and magnesium. Therefore, not surprisingly, the color of shale is very variable. Black shales, for example, are due to their carbon content; red shales indicate the presence of hematite; browns, and yellows come from differing proportions of goethite and limonite. It's interesting that these color differences have failed to attract any of the reverence accorded to analogous variations in the related rock slate (Chapter 6).

The Dissolved Component Makes Rocks

The deposits from evaporating seas and lakes mentioned earlier, evaporites, can harden into rock, such as the gypsum at Steigerwald in Franconia, Germany. (The strata below the town of Burton in the English Midlands are rich in evaporitic gypsum, producing well-water that gave character to the Export and India Pale Ales that originated there.) Limestone can also be an evaporite, where the rock has formed by water evaporating. Examples occur in parts of Corbières, Languedoc, and some of the limestone bedrock at Chateau Margaux was precipitated from a freshwater lake.

Now we come to **tufa**, a term frequently misunderstood in the wine world (and elsewhere). Geologically, it refers to calcium carbonate deposited locally from cold water, such as in waterfalls, cold springs, and seeps. The confusion takes us back to those master masons the Romans, who prized their local rocks for the monumental buildings that befitted their capital. They called the rocks *tofus* (or *tophus*), neither knowing nor caring that they had differing origins. With time, the name became extended throughout the Empire for any kind of softish, porous rock that made good building blocks. Eventually, it became known as *tuffe* in French and as tuff in English. The name held sway in England until around the eighteenth century, when the stone dealers of the day learned that Italians were now calling it *tufa* and convinced architects that this exotic-sounding material was preferable to old-fashioned, Anglo-Saxon-sounding tuff. So both tuff and tufa became used for any kind of rock, irrespective of its origin, as long as one could build conveniently with it.

But when geologists saw that quite different substances were being given the same name, an agreement evolved that the name "tuff" should be restricted to volcanic material. Consequently, for well over a century now, geologists have used tuff to refer only to volcanic rocks, and tufa to a calcareous rock formed by localized precipitation from cold groundwater. However, outside geology, the lack of distinction persists. For example, many Italian vineyards with siliceous soils are reported in English as being located on tufa, even in markedly volcanic areas such as Soave and Campania—home of the original tuff—while tuff is used in calcareous areas such as Piemonte, Puglia, and parts of Tuscany. As mentioned in Chapter 3, there are extensive deposits of tuff in northern Hungary. Numerous wine cellars have been excavated in the material, and some show conspicuous volcanic fragments in the walls. Yet in the publicity material of a prominent Eger winery, a leading travel guide, and throughout at least three books in English on Hungarian wine, this material is incorrectly referred to as tufa.

In geology, then, tufa is precipitated from cold water. More relevant for vineyards than localised mounds of tufa around a single spring are the sheet-like occurrences that can cover substantial areas. For example, tufas can be precipitated from rivers, especially where there is limestone or other calcareous material in the vicinity, as happens in the Ebro and Duero valleys in Spain. They can also arise from lines of springs spread along hillsides. This happens in the Val di Noto, Sicily, where wine caves have been dug into the soft calcareous rock (but called by some guidebooks "volcanic tufa"). Tufa deposits are known around the world; some 500 sites have been noted in Hungary alone. A number of U.S. states have tufa deposits: it forms in some upstate New York lakes, for example, and in Virginian rivers along the Eastern Appalachians, including the Shenandoah Valley AVA.

There is another twist to this confusion. In the middle Loire region of France, including the eastern Anjou and Touraine AOC areas, there is a sandy marine limestone, much used there for caves in which wine is stored, mushrooms are grown and, until not long ago, humans lived. The rock was quarried for building celebrated Loire chateaux such as Chambord, Blois, Chenonceau and Amboise. It floors vineyards around Saumur and Chinon, and so it sometimes appears in English wine writings, under its local French name: *tuffeau*. But one major book on French wine calls it tufa. More confusion!

Tuffeau is easily seen in cliffs along the Loire River, say between Tours and Saumur, yellowy in color due to the rock's constituent glauconite having weathered to ochrous limonite. It contains abundant fossils but a characteristic here is that many are broken up, the result of frequent storms disturbing the shallow seas at that time. I mention this in order to emphasise that tuffeau is a sedimentary rock, areally extensive and stratified in layers due to accumulations through time of submarine debris. It is not a chemical precipitate. Thus, although the word tuffeau may sound like tufa, the nature of the two materials and their geological origins are different: *tuffeau is not a synonym for tufa.*

The historic town of Orvieto, Italy, is well known to tourists for its photogenic hilltop setting, which supposedly has something like 1200 caves excavated into it, and to wine lovers as one of the better known Umbrian wines. The rock is a tuff, rather soft because it is geologically young, having been ejected from a nearby volcano about 315,000 years ago. But, as in the Hungary example, some guidebooks to the region call the material tufa, and at least one wine book likens it to the rocks of the Loire and calls it tuffeau. In other words, here we have three words that look rather similar and are sometimes used as though they all mean the same thing. Geologically, however, these three materials are quite distinct: tuff is volcanic, tufa is a calcareous precipitate, and tuffeau is a marine limestone in the Loire region of France.

I should mention a further "t word", the warm water equivalent of tufa: **travertine**. Because of the greater amounts of calcium carbonate that can be dissolved in hotter water, substantial thicknesses of travertine precipitate can accumulate. The warm waters also allow a lot of carbon dioxide to be dissolved in them, which on cooling comes out as a gas, giving the irregular cavities and holes that are typical of much travertine.

Pitigliano in Tuscany, Italy, is well known for its extensive deposits of tuff (though some English tourist literature on the region trumpets it as the "Land of Tufa" and the "Queen of the Tufa Towns") but over 15 square kilometres of the Bianco di Pitigliano DOC is underlain by travertine, which, being calcareous, gives alkaline soils that contrast with the acid soils derived from tuff. At Bagni San Filippo, in the Val d'Orcia DOC, the travertine masses are over 40 meters thick and still active: several centimetres have to be scraped away each day from the swimming pool in the local spa. The name travertine, or its equivalent in other languages, appears on wine labels from Abruzzo's Colline Pescaresi and Tuscany's Grance Senesi and Monteregio, and from Pouilly Fumé and Saint Pourçain-sur-Sioule in the Auvergne; it's the name of a winery in Australia's Hunter Valley.

Even so, travertine is probably more familiar as a building stone: creamy white and polished, with irregular dark holes. Many a grand building in Europe (visualise the shining white of Sacré-Coeur in Paris) and North America (such as New York's Pennsylvania Station and Chicago's Union Station) are at least faced with travertine. Mediterranean towns were built with it, such as the Italian Marche's Ascoli Piceno or Turkey's Pamukkale. You may have slipped on the shiny, shoe-buffed streets of towns such as Rovinj or Dubrovnik, paved with gleaming travertine. The quarries of Rome have been producing travertine since classical times: one modern author calculated that 100,000 tons of it were extracted just for Rome's Coliseum. Later, Michelangelo used it to face St. Peter's Basilica and its dome; Bernini used travertine for the splendid Colonnades that arc around St. Peter's Square. Today, destinations for the rock are less rarified: modern Roman quarrymen toil to produce facing stone for a well-known chain of restaurants with golden arches.

The Biology Input and the World of Limestone

On burial, oceanic siliceous organisms such as radiolaria become reorganized internally and compacted to produce flint (discussed in Chapter 3). Diatoms lithify to produce the siliceous powdery rock known as **diatomite** or **diatomaceous earth**. This material was first recognized in Lüneburg Heath in north Germany and was called *kieselguhr*, a word commonly used in the wine industry today. Actually, it has a remarkable range of uses, from cat litter through poultry feed to dynamite, but with wine it's used mainly as a filtering medium. The sheets used to carry out filtration are pressed cellulosic fiber impregnated with kieselgur. The round, tightly packed diatom particles provide a high permeability, but their fineness traps the minuscule particles that would otherwise pass through or clog filter paper. At the same time, being made of inert silica, kieselguhr does not interfere with the wine's color or flavor. Some growers spread diatomite on vineyard soils in an effort to improve their permeability and resist compaction (Chapter 9).

The fineness of diatomite can allow any fossils in it to show exquisite detail, such as the plant leaves at Monte Amiate, in Tuscany, Italy. The oldest known records of grape vines are the leaves preserved in the 10-million-year-old diatomite in Kisatibi, near Akhaltsikhe in Georgia. Diatomite in the Santa Rita Hills AVA, northwest of Santa Barbara, California, provides well drained slopes prized by the growers there; one local winery is called simply "Diatom." The world's largest commercial production of diatomite is in the adjacent area, around the town of Lompoc. The movie "*Sideways*", much loved by many a wine lover, was filmed in this region: the dazzling white road cuts and bluffs along the Santa Ynez River that the protagonists Miles and Jack see as they carouse at the Sanford winery are striking outcrops of diatomite.

And so to limestone. Being made primarily of calcite, limestone is very different from the siliceous rocks and, generally speaking, has certain consistent properties. It weathers readily (Chapter 9), which gives rise to characteristic landforms (Chapter 8) and to well-drained, high-pH soils (Chapter 10). That said, limestone is much more variable than many wine writings imply. It's very variable in appearance, with various combinations of cemented calcareous debris and fine calcium carbonate that was chemically precipitated or secreted by organisms. Pure calcite limestones are pale gray to white—and rare because almost always there are impurities that give a coloration. On the one hand, in coarse-grained limestone, the cleavage surfaces of the calcite may make the rock glitter in the sunshine; on the other hand, the calcite may be very fine-grained, giving the rock a dull, featureless appearance, breaking in a curving fashion because there is nothing for the fracture to follow.

Some limestones are conspicuously rich in fossils (Figure 5.11; see Plate 7). If the kind of organism can be identified and is abundant, it can lend its name to the rock, as in **crinoidal limestone** and coral-packed **reef limestone**. The name **coquina** appears on some wine labels, such as those from Hawkes Bay, New Zealand, and Arroyo Grande Valley AVA, within the Central Coast AVA of California. This is a poorly cemented limestone made almost entirely of fragmented but coarse-grained marine shells and skeletons. Oolitic limestone is composed almost entirely of ooids (see above) cemented together. It occurs in a number of vineyards on the Côte d'Or and in Alsace, as well as in central England, where the rock makes a swath across the country, underlying several vineyards and providing a famous architectural stone. Iron carbonate minerals such as siderite are often present, even dominant. Such **oolitic ironstones** occur in some Burgundy vineyards, for example, at Savigny-lès-Beaune, and elsewhere it can be a commercially important source of iron.

The magnesium-rich limestone called **dolomite** (Chapter 2), composed mainly of the eponymous mineral, is only seen forming today in specialized environments such as highly saline lagoons and apparently in the presence of certain bacteria. But it is common in ancient sedimentary rocks, for reasons that are not well understood. The current thinking is that dolomite rock (sometimes called **dolostone**) is partly the result of waters rich in magnesium percolating through limestone and gradually replacing some of the calcium and partly the result of bacteria helping to precipitate dolomite on the ancient seafloor.

Then there is that rather charismatic limestone much loved by wine people: chalk. The difficulty here is that the word has a number of meanings, not all of which are very precise. Geologically, **chalk** is a specialized form of fairly pure limestone, composed largely of minute fragments of coccoliths (see the earlier Contribution from Biology section and Figure 5.7). There may be some foraminifera in the rock, which are also calcareous. Fragments of sponges, diatoms, and radiolaria (all siliceous), together with a small proportion of clay minerals such as glauconite, can contribute a small proportion of silica. But the rock has to consist primarily of calcareous microorganisms, especially coccoliths. It's highly porous and tends to be crumbly, but it's not necessarily as weak and soft as is sometimes made out; after all, it stands proudly against the waves in English headlands such as the White Cliffs of Dover and where Beachy Head, in the words of Charlotte Smith (1749–1806), "rears its rugged brow above the channel wave."

Chalk in its proper geological sense extends from the northeast coast of Yorkshire southward through England, in places covered by younger strata, under the English Channel to France and then patchily across Europe on to the Judean Desert and the hills of Crimea and Kazakhstan. It also occurs in Texas and parts of some other states in the United States. Chalk famously underlies some of the vineyards of southern England and of Champagne, and some English growers like to maximize this, stating

Figure 5.11 Limestones containing fossils: (a) Limestone landscape and loose blocks near Postira, Brač, Croatia, in which a depression in the ground provides some protection to the small vineyard from the island's winds; the inset is a close-up of the rock. (b) Carboniferous limestone from northern England.

that "our soils are identical to those in Champagne"; "our iconic soil, the same as that of Champagne"; and the like. However, the rock varies within Champagne (Chapter 12), and in England its variations lead geologists to divide the Chalk formation (a collection of chalk strata and other sedimentary rocks, Chapter 3) into no fewer than eight subunits. In other words, even within its true, coccolith-rich meaning, chalk can present a variety of appearances and properties.

Usually though, chalk makes an excellent substrate for viticulture, with its restrained nutrient provision and high porosity providing excellent water storage coupled with good permeability. After all, it's the most important aquifer in northern France and southern England. However, although these water properties are often ascribed to pores between the constituent calcite grains, they're primarily due to microfissures in the rock—it's an excellent example of fracture porosity and permability (Chapter 10).

The special sea conditions needed for this deposit to form arose in a period of geologic time known as the Cretaceous (from the Latin word for chalk), and this is where confusion sets in. In western Europe in particular, sundry calcareous deposits are referred to as chalk if they formed in Cretaceous times. For example, the soils of Cognac (France) and the *albarizas* soils of Jerez (Spain) are often called chalk, but they don't fit exactly with the geological meaning. Indeed, in some wine writings, chalk is used for any vaguely fossil-bearing, crumbly limestone. Chalky is sometimes taken as a synonym for calcareous and sometimes even as simply another name for limestone. If the word "chalk" is to have any use in understanding vineyard geology, it has to be one particular kind of limestone.

So, chalk is one kind, a subset, of limestone. The two terms are not synonymous. There are further issues: calcium carbonate has no flavor, yet wines are commonly described as having a chalky taste. I've heard various proposals about what this actually means, but perhaps most often it is said to express a mouthfeel rather than a flavor. Presumably, this has to do with a perceived dustiness, perhaps from an assumption that all chalk is crumbly or maybe for some reviving memories of classrooms dusty with blackboard chalk. But then schoolroom chalk is made of gypsum, not calcite. The famous "chalk dust" of disputed tennis line-calls is titanium dioxide, and tailors' chalk is made from the mineral talc. So, the word "chalk" is not straightforward!

Another troublesome word is **marl**. It's an old Saxon word and not much used in geology these days, though wine writers seem to like the term (one 2016 book on wine considers limestone just a form of marl). Essentially, it's a rock with a composition somewhere between a fine-grained limestone and a mudstone. Efforts have been made to define the proportions with some rigor, but these tend to be obscured by the time-honored usages that are still around. Thus, although in a modern scheme the limestone–mudstone proportion is defined as 50–50, plus or minus 15%, some rocks from Ontario that continue to be called marls have around 80% limestone, whereas others from England and from southern Germany, where the term originated, have as little as 20%. Another problem, in addition to these entrenched names, is that given the very fine-grained nature of the rock, it's difficult to tell the proportions without analytical equipment.

All this haziness is compounded in the wine world by a variety of the celebrated rocks and soils of the Côte d'Or being described simply as marl. One frequent misconception is that marl is a *mixture,* with visible pieces of limestone interspersed with clay. But marl is very fine grained; it contains clay minerals and calcite that is far too fine to be differentiated, in the same way that we see gray as a color and not as a mixture of black and white. A representative marl looks like a pale-colored claystone, wholly featureless but breaking with a conchoidal fracture, and to some extent fizzing with dilute acid. But some marls have very little silt and so are limestone-claystone intermediates, with a slight fissility like shale. As a final twist, technically speaking, marl is a sediment (Figure 5.6); the rock should properly be called a **marlstone**.

Most limestones, then, formed in warm, shallow seas. It just so happens that around 200 million years ago, plate tectonic processes produced just such conditions in the area that eventually was to become southern Europe, leading to the deposits that form much of the land surface today (Figure 5.12). So when the Romans came to this part of the world and encouraged vine growing, in places like Burgundy, the Loire Valley, Cognac, and Champagne, there happened to be limestone in the ground in which the vines were planted. In these climatic conditions, the limestone soils suited the vines well, and the rest is history. But in the same way that it's interesting to speculate on what the world would be like if geology had bestowed, say, rich oil deposits on the Horn of Africa instead of the Middle East, we can wonder how vineyard geology might be viewed if France wasn't substantially underlain by limestone. It's an intriguing vignette of the philosopher Will Durant's aphorism: "civilization exists by geological consent."

Sedimentary Rocks Come in Layers

Try to imagine the seafloor just out from a river estuary, with sediment billowing in from the river and slowly settling. Then something changes. The sediment is suddenly different, in particle size perhaps or in color. Possibly the seashore currents have changed, or the river's upstream course has moved so that it is now transporting different material; perhaps there was a recent storm, and the river is flowing faster and is even in spate, so that its sediment load is suddenly very different. The point is that because the incoming sediment keeps changing in pulses, its accumulation on the seafloor will be layered.

Then, when it all becomes lithified, and if the seafloor is uplifted to make land so that we can now see the material, we will see this layering. The time-honored geological term for it is **stratification**, and the layers are **strata** (though the days of

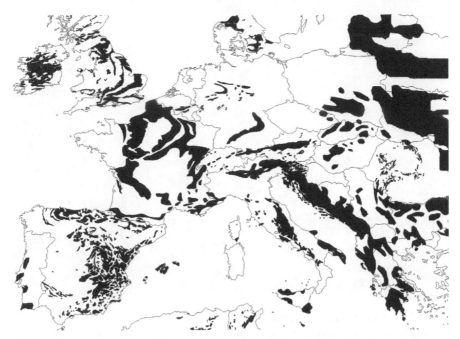

Figure 5.12 The distribution of limestone at the land surface in Europe. Adapted from the *World Map of Carbonate Rock Outcrops*, vol. 3.0, University of Auckland, New Zealand, at http://web.env.auckland.ac.nz/our_research/karst.

speaking about an individual stratum are long gone). Most often though, geologists talk about **bedding**, and each layer is a **bed**. The thicknesses of individual beds can be anywhere from a few millimeters up to, unusually, many meters.

This bedding is the hallmark of sedimentary rocks (Figure 5.13; see Plate 8). There are unusual sedimentary materials that are unstratified—tillite and loess, for example—but even there the deposit as a whole is a layer, albeit an irregular and very thick one. Stand in one of the vineyards around Stellenbosch, South Africa, and look up to the massive hulk of the Simonsberg. About a third of the way up, the vegetated cones of debris give way to craggy bare-rock faces in which bedding is clearly visible. You see colored bands slightly losing altitude toward the south. Not far to the north is another isolated rock mass, but its crags show nothing like this: it's the Paarlberg, and it's granite.

Beds of sedimentary rock commonly change in thickness as we trace them along a cliff face. It's because sedimentation on the seafloor is usually pretty irregular, as pictured earlier for the Bahamas, and Chapter 8 describes how deposition on land is particularly erratic. A sediment deposited in a channel, for example, below the sea or in a river, will go very quickly from nothing at the margins to a maximum in the center of the channel and back to nothing again. Geologists talk of lenses of sediment to express such localized deposits. The lenses of water-holding

Figure 5.13 Stratification or bedding, the hallmark of most sedimentary rocks, here inclined from its original horizontal orientation downward toward the left (north) of this picture. Near the Cederberg winery, Western Cape, South Africa.

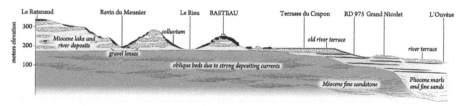

Figure 5.14 Sketch cross section across the Rasteau AOC district (southern Rhône, France) showing the irregular nature of bedding. Based on a figure by Georges Truc at http://www.dico-du-vin.com/rasteau-cotes-du-rhone-meridionales.

clay within the gravel mounds of the Médoc are famous, and Figure 5.14 sketches another example.

The underlying point of all this is that vineyard geology is more intricate and variable than we often imagine. Sedimentary rocks and processes may be relatively easy for us to visualize and comprehend, but they can be capricious. We might read of a particular vineyard being "on limestone," but we have seen that limestone is variable in chemistry and in physical properties, and it weathers to give a similarly varied collection of soils. In other words, the phrase doesn't convey very much, even where only one rock name is involved.

So, we live on the surface of a planet that is very mobile, with complex processes operating on it. Will things be simpler if we go down into the Earth's solid crust? It's not so. We will see in the next chapter that mobility and change are not confined to Earth's surface.

Further Reading

Jones, Stuart. *Introducing Sedimentology.* Tampa, FL: Dunedin Press, 2015.

Stow, Dorrik A. V. *Sedimentary Rocks in the Field: A Colour Guide.* Boca Raton, FL: CRC Press, 2005. This work is aimed at students and contains material that is not of obvious relevance to vineyards, but it has lots of photographs of sedimentary rocks.

6

Metamorphic Rocks

The Drive for Change

We come now to the metamorphic rocks, the result of modifications to already existing rock. I'm well aware that this can all seem a bit mysterious. After all, no one has ever seen the changes take place; no one has ever witnessed a metamorphic rock form—the processes are imperceptibly slow, and they happen deep in the Earth's crust, way out of sight. Why should these changes happen? Well, they are primarily driven by increases in pressure and temperature, so we begin with a look at these two factors.

There are sites in the Earth's crust where material becomes progressively buried. It happens, for example, where a tectonic plate is driving underneath another one, taking rocks ever deeper as it descends. It can happen in the central area of a plate that is stretching and sagging, allowing thick accumulations of sediment. It's pretty self-evident that as buried material gets deeper, because of the growing weight of rocks above bearing down due to gravity, it becomes subjected to increasing **burial pressure**.

Less intuitive, though, is the fact that this pressure acts on a volume of rock equally in all directions. Imagine a small volume of rock at depth. It's bearing the weight of the rocks above it, and so it responds by trying to move downward and to spread out laterally. Of course, it can't because it's constrained all around by other volumes of rock that are trying to do exactly the same thing. And so the downward gravity is translated into an all-around pressure. It's the same effect as diving down to the bottom of a swimming pool. You feel the increased pressure owing to the weight of water above, but you feel it equally in all directions. All-round pressure like this can cause things to change in volume, through changing their density, but it can't change their shape.

However, there can be another kind of pressure as well, and this does have direction, and it can cause change of shape. In the Earth, we call it **tectonic stress**. It comes about through heat-driven motions in the Earth, including the movement of tectonic plates. Its intensity varies from place to place and through time—it waxes and wanes; it may or may not be acting on a particular rock. But by having a directional aspect, it can cause rocks to change their overall external shape, which we'll look at in the next chapter, and to change their shape internally, which is what concerns us here. Continuing the swimming pool analogy, if you dive with sufficient energy you may hit the bottom of the pool, and that directional stress is capable of changing your shape, causing damage.

Material that is being buried will at the same time be subjected to increases in **temperature.** As we saw in Chapter 1, this is largely because minerals undergo radioactive decay. For much of the Earth's crust, the rate of increase is around 25°C per kilometer of depth. Geologists usually reckon that significant changes in rocks start at around 200°C and mainly happen in the 300–600°C range, so with regard to producing metamorphic rocks, we're talking about depths of around 12 to 24 kilometers. These may sound like enormous numbers, but they're pretty modest for the Earth as a whole. In fact, if the depths were much greater, then the material would be unlikely to ever become visible at the Earth's surface.

It's not uncommon in popular literature to read that metamorphic rocks form at profound depths, down in the Earth's core, and at unimaginably huge temperatures, even having been melted. However, the changes take place mostly within the Earth's crust and by no means too deep for the results to be exhumed through time. And with the one exception discussed in the next section, no melting is involved. *The changes take place in solid rock.* It's reminiscent of that children's toy with letters printed on plastic buttons that can be slid around to form new words, utilizing the one space that lacks a button. The order of the letters changes, but the toy remains intact. As for the mind-boggling temperatures, well, a hot kitchen oven generates temperatures approaching the realms of metamorphism. It's the slowness of the changes that is incomprehensibly remote, as these solid-state processes take millions of years and more. (So there's little point in trying to reproduce them in your pizza oven.)

So burial pressure and temperature increase with depth below the ground, but we're still left with the question of why the rocks change. The answer is that it's all a matter of their trying to maintain stability in the changing conditions. A crystal lattice with its locked-in ions has internal energies, and these have to be at a minimum for the mineral to be stable. If the ambient conditions change, there may be an arrangement that gives a lower-energy state, and the crystal will try to achieve it. The increasing heat and pressure provide the energy needed to drive the work. Then the results become locked in if the rock is uplifted and excavated, as the energies needed to drive any further changes drain away. Similarly, a dough made of flour and water is pretty stable at room temperature, but in an oven at 200°C or so it will change! Then as things cool those changes will remain—just as well for bread eaters.

If the changes take place in the presence of a tectonic stress, then any platy-shaped minerals will be induced to adopt a new physical orientation at right angles to the stress, in order to maintain physical stability. Think of a pack of cards; each card has an extremely flat shape. They could be arranged in a "house-of-cards" structure, although this is not at all stable physically. Certainly if there was a small input of energy (namely, you jostle it), the cards would strive to attain a more stable arrangement; they would collapse, all aligned at right angles to the directional stress, which in this case is gravity. A number of metamorphic rocks develop a new "planar aspect" such as parallel-aligned minerals, and to cover the various forms, there's a very useful word: **foliation**. As we shall see shortly, some metamorphic rocks are **foliated**, some are not.

We bring together all these changes—in the kinds of minerals, their size, and their physical arrangement—under the term **metamorphism**. It's worth saying it out loud; undergraduate students are afraid of the word and like to put in extra syllables. Part of the problem is that the word "metamorphosis" is better known. Moreover (according to Google), there is also the word "metamorphosism." In geology, however, it's metamorphism, and the verb is to **metamorphose** a rock.

Figure 6.1 outlines a simple scheme showing the names of the metamorphic rocks most often seen in wine writings. It's very generalized. Importantly, all the boundaries between the divisions are gradational: for example, the planar element

Original rock	Metamorphic rock
Mudstone	*FOLIATED (has a planar aspect)*
Siltstone	**SLATE** (very fine, splits well)
Shale	**SCHIST** (visible aligned minerals)
Volcanic tuff	⟋ORTHOGNEISS (igneous origin)
Igneous rocks	**GNEISS** (BANDED) ⟍PARAGNEISS (sedimentary origin)
	NONFOLIATED
Limestone	**MARBLE**
Quartz-sandstone	**QUARTZITE**
Gabbro	**AMPHIBOLITE**
Peridotite	**SERPENTINITE**
Any original rock	**HORNFELS** (featureless, due to heat) **MYLONITE** (highly stretched in fault) **MIGMATITE** (partially melted: pale and dark zones and patches)

Figure 6.1 A classification of metamorphic rocks, showing the names used in the text.

in some foliated rocks can be very weak; conversely, rocks here called nonfoliated can occasionally have a distinct planar look.

The Diverse Dynasty of Slate, Schist, and Gneiss

Mudstones and shales are normally the first materials to become unstable on burial because they are both full of clay minerals, which are sensitive to change. The clays reorganize into slightly denser, more stable sheet minerals such as chlorite and mica, and if a tectonic stress is operating, they take on a parallel alignment. As a result, the rock will take on a new, distinct tendency to split, which a microscopic view shows is controlled by this new mineral alignment. We call this particular kind of foliation a **rock cleavage**, and the rock that shows it par excellence is **slate** (Figure 6.2; see Plate 9). Crucially, the orientation of the cleavage depends on the orientation of the tectonic stress that caused it and so it bears no relation to bedding of the parent

Figure 6.2 Photographs of slate. (a) The slaty soil at Cederberg winery, Western Cape, South Africa. (b) Roofing slate being split in the time-honored way in North Wales, United Kingdom. (c) A block of the same slate, here with the cleavage running approximately horizontal. (d) This is the same slate viewed down a microscope to show the intense alignment of fine-grained platy minerals (the field of view is about 3 millimeters).

sedimentary rock. Thus, it contrasts with the fissility of a shale, which always paral-
lels bedding (Chapter 5).

In its early stages, the new mineral alignment may be weak, and sedimentary bedding may still be sufficiently present to hinder a clean, planar splitting. Thus, there is
no clear boundary between shale and slate, which is why some reports on vineyards
in the Rheingau, for example, say they have shale soils while other literature declares
that they have slate. Neither is wrong. The middle Mosel is similar in that it's a classic area of slate quarrying and of famously slaty vineyards. Yet some of the rocks are
celebrated for their beautiful, delicately preserved fossils, even though metamorphism normally destroys any fossils. In other words, some of these Mosel rocks are
in this intermediate shale–slate "no-man's land." Slate with significant amounts of
quartz cleaves less well and grades into a fine-grained **foliated quartzite**.

The durability, low permeability, and porosity, not to mention insolubility, of
slate makes it arguably the finest roofing material available and a very attractive
material for kitchen worktops. Incidentally, the word "tile" is properly used for
a manufactured product, which is normally some sort of fired clay. *Shingle* is all-
inclusive. A distant view of many a European town or village is characterized by its
roofs, composed perhaps of dark slate or of bright red tiles, according to whether
slate or clay was available locally.

Slate is found in numerous wine-producing areas, including the Clare Valley,
South Australia; around Cebreros in the south of Castilla y León, Spain; Kitzeck, in
the South Styrian region of Sausal, Austria; central Chile's Aconcagua Valley; and
the Huan Hills vineyards of Thailand. It's common near Angers, just north of the
River Loire in France, where the constituent minerals in some slates have a natu-
ral elongate shape, making the rocks break into more of a rod shape than a sheet.
This tendency is exploited to make slate stakes, fence posts, and trellis supports
(Figure 6.3).

Most slates are gray in color—Wordsworth nicely called it "hoary grey"—but
there is a whole range of subtle hues. Slate colors have found their way into wine
names, and in the quarrying trade they are given charming names such as royal purple, sea green, and heather red. The Camel Valley vineyard in Cornwall, England, is
located on a gray slate bedrock, which nearby is actively quarried for a whole range
of slate products; Pant Du vineyard in North Wales uses as a mulch plum-colored
slate waste from a quarry just a couple of kilometers away. In Valdeorras, northwest
Spain, the spectacularly terraced vineyards above the River Sil have a distinctly slaty
soil, and nearby the bedrock is quarried for the smooth, rather shiny, dark gray slates
that are seen on roofs all over the world.

So is there something special about slate for wine tastes? The belief is probably
nowhere more embedded than in the vineyards alongside German rivers such as
the Mosel and Rhine. Certainly, any wine taster seeing those strikingly steep slopes
and vines growing in what seems to be no more than slate rubble will be expecting
something distinctive. And the wines are, of course, just that, but the idea that the

Figure 6.3 Slate with constituent minerals that are more elongate than platy tends to break into rods, which geologists call a "pencil slate." Such a material is used here as straining posts, in the Muscadet region of the Loire, France.

slate can actually be tasted in the wine, as some claim, has to be metaphorical. Vine roots cannot take up the cleavable bonded aggregates of geological minerals that make slate what it is (see Chapter 9), let alone somehow transmit them so that slate exists in the finished wine. In any case, like most rocks, slate lacks any taste or odor. To have a taste, a substance has to dissolve, and manifestly that is not the case with an inert material that makes practicable kitchen countertops and durable roofs. The current fashion in some restaurants where food is served on slate "plates" would lead to a different experience if the taste was being affected!

In cool, northern latitudes like Germany, the chief physical factors influencing the grapes are the climatic effects arising from differing gradients and aspects of the hillslopes and from the proximity to a river. The main contribution of the slate soil itself is its free drainage, which is so important in a moist climate. At the same time, the interlocking, jagged slate fragments give the soil a cohesion that helps reduce erosion on these remarkably steep slopes. Also, the dark colors of the slates may contribute some warming effect and the smooth cleavage surfaces might produce some light reflectivity (Chapter 10); both of these effects are possibly of significance in climates where the grapes are striving to ripen. The chemistry appears relatively unimportant in that most slates weather to yield the full range of nutrients required by vines.

There is, however, the intriguing matter of reported differences in wines produced from vines growing near each other but on soils with different-colored slate. For example, high in the empty and spectacular Cederberg Mountains of the Western Cape, South Africa, a difference is perceived between wines from Shiraz vines growing in red slate and those growing nearby but in gray slate. Riesling wines from the blue slates of Wehlen, in the middle Mosel, Germany, are promoted as being different to those from the red slates there. So what could be causing this difference? The thermal properties of the soils presumably differ slightly, but this would hardly seem to be significant in those parts of the world with an adequately warm climate. Some research suggests that the wavelength of the reflected light, which could vary somewhat with slate color, can affect grapes. In other respects, however, the physical properties of the slates are identical: a single slab of slate can show different colors but is still physically homogeneous. And the obvious candidate, the chemical differences that are responsible for the colors, doesn't really hold up.

The different gray colors of slates are due to variable amounts of carbon that is spread very finely through the rock. These variations can scarcely affect the vines, though, because the amounts are tiny and, of course, plants get their carbon from the atmosphere through photosynthesis. Other color variations depend on minute differences in trace elements such as titanium, vanadium, and chromium; again these are unlikely to affect vines as the roots are barely able to take up these non-essential, potentially toxic elements, which are present in wine in such minuscule amounts that they are not normally reported. The chief color influence, certainly for red hues, is the amount of iron in its ferric form, Fe^{3+}. But this is insoluble and hence unusable directly by vines; the small amount of iron that vines need is absorbed in its soluble, ferrous form, Fe^{2+}. Perhaps it isn't the visible geological differences that are relevant but the other factors that will be changing along with them, such as those having to do with mesoclimate and microbiology. Otherwise, a geological explanation remains elusive.

So what happens if the slate isn't uplifted and exhumed but continues to be buried? The ambient temperatures and burial pressures continue to increase, acting to enhance the processes of forming new minerals and enabling them to grow to a larger size. If this material does eventually find itself at the Earth's surface, the minerals will have become big enough for us to see, and the platy ones will clearly show their alignment. But perhaps surprisingly, with these larger particles this rock will no longer break as easily and cleanly as a slate. In other words, we now have a different kind of rock, one known by the ancient name of schist.

Once again it's a gradational division: different geologists "draw the line" in different places, such that for one geologist a vineyard may be sited on slate, while another would call the rock a schist. Some geologists interpose another rock name here—**phyllite**—but this hardly rids us of the difficulty. Moreover, in other European languages, the equivalent terms have different ranges of meaning (e.g., the

German word *Schiefer* is commonly used for both slate and schist, as is the French word *schiste*, which is also sometimes extended to include shale).

The gradational nature of the boundary between slate and schist is well illustrated in the Douro region of Portugal and in Priorat, Spain. For both areas, some accounts report a schist bedrock, while others call it a slate. Again, neither is wrong. For the Douro vineyards, the important fact is that the foliated rock—there is no difference of judgment on that—is so oriented that the planes of weakness resulting from the foliation are steeply inclined, allowing both rainwater percolation in wet times and penetration by vine roots.

In the scheme presented here, a **schist** is defined by a parallel alignment of minerals that are themselves clearly visible to the unaided eye (Figure 6.4; see Plate 10). This arrangement is termed a **schistosity**. Various amounts of quartz can be present either in thin bands or among the aligned minerals, but because of its natural roundish shape does not contribute to the schistosity. So, some schists that lack quartz shatter finely, whereas some quartz-rich examples break into large slabs and splinters. At Morgon in Beaujolais, the schist-derived, dull red soil is distinctly crumbly

Figure 6.4 Photographs of schist. (a), (c), and (d) The silvery sheen often imparted by aligned micas. (a) and (d) also show dark minerals newly formed during the metamorphism; (c) shows the fine crinkling, properly called crenulation, that is common in mica schists. (b) A view down a microscope to show intensely aligned plates of mica, giving a schistosity in this view running from northwest to southeast. The field of view is about 3 millimeters: contrast the size of the minerals with those in a slate, as in Figure 6.2d.

(the locals call it "rotted rock"), and at Cascastel-des-Corbières in the Languedoc, the tiny fragments of schist make a black vineyard soil that trickles between your fingers. In northern Corsica, by contrast, some of the schists are much richer in quartz, and the rock breaks into substantial slabs that are used to build houses.

Although schist does not split as cleanly as a good slate, because of the coarser grain size, it can make an attractive decorative stone. Nowhere has the rough splitting of schist been utilized more effectively than along the western seaboard of southwest Scotland and Ireland, where slabs of the local chlorite schist were used in premedieval times to carve the iconic, pierced Celtic crosses that are replicated on many a Christian gravestone and in jewelry. Even the word "schist" itself is of ancient origin. The Greek lapidary Damigeron, writing in the second century B.C., recommended mixing *ischistos* with "women's milk" to counter "phantasms and hallucinations" and various other inscrutable complaints, while Pliny the Elder noted in his great compendium "Natural History" that *schistus*, whose "nature thereof is to cleave along into certaine filaments or threads like haires," was good for such things as bloodshot eyes, "female discharges," and bladder problems. Wonderful stuff, schist. And in 620, Archbishop Isidore of Seville was remarking on how easily the rock he called *schistos* broke and how it "gleamed in the light, just like saffron", he said.

A wide range of minerals can contribute to schistosity, providing they don't have roundish shape. Schists are therefore varied in appearance, including color. Small amounts of graphite, originating from organic carbon in the parent rock, are commonly dispersed through the rock, giving a dark gray or even a black color. Where the schistosity comprises aligned muscovite, the rock can gleam as the silvery plates reflect sunlight. A particular amphibole (called glaucophane) gives a distinctly blue tinge to some schists in the Coteaux du Cap Corse district of northernmost Corsica. Biotite gives a brown appearance and green schists are common because of chlorite and epidote.

In other words, the term *schist* covers a wide range of minerals and chemistry. All they have in common is some degree of schistosity. Yet the dominant thinking about wines produced on schist, in contrast to those produced on slate, is to emphasize a commonality. A case in point is Faugères, in the Languedoc, France. Almost the entire AOC area is on schist, which the publicity claims is the key to the personality of its wines, even though the rock is highly variable and yields soils composed of anything from crumbly shards through slabs that can be used for building. Some of these soils are far too acid for viticulture.

Despite this variability, there is an organized consortium of wine producers—L'Association Terroirs de Schistes—which maintains that their wines are distinctive because their vineyards are located on schist, even though they include such diverse places as Kastelberg in Alsace, Côte-Rôtie, Savennières, Collioure, St-Chinian, Banyuls, Coteaux du Cap Corse, and Valais in Switzerland. Notwithstanding the disparity in physical conditions between these regions, the markedly different cultivars, styles of wine, viticultural and winemaking methods, a study commissioned in 2013 from the Agricultural Institute in Montpellier (SupAgro) by the consortium concluded that schist bestows (in unspecified ways) a discernible character on its members' wines. What is this distinguishing hallmark? The report says that it's "a nice balance of freshness, nice acidity, supple and silky tannins," with "minerality and salinity being the most resilient feature." Hmmm.

The variability of schist is also illustrated by the viticultural region with which in recent years it has perhaps become most closely associated: Central Otago, toward the south of South Island, New Zealand. Much of the material here is silvery gray, but it varies from being quartz-rich and strong, and easily extracted in slabs for building purposes, to crumbly, delicate schists rich in mica. While most of these were originally muddy sedimentary rocks, some schists in the area were originally volcanic ash and are distinctly green in color because of their chlorite content. Hence, although it's fashionable in some descriptions to remark that the wines are made from "grapes grown on schist," this observation can't convey much, as schists are so variable.

Now for some fine semantics. The adjective for a rock showing a schistosity is **schistose**. Thus, geologists may talk of, say, a quartzite being slightly schistose. Early geologists used the now obsolete word "schistus," though for rocks different from schist. However, wine journalists seem to have invented a new word: **schistous**. It's used to describe vineyard soils, and it seems to mean that the soils consist of fragments of schist. The word doesn't exist in the geological lexicon, but neither does an equivalent term. So I acknowledge it here as serving a useful, legitimate function. By the same token, it's incorrect to describe a soil as schistose; a soil cannot possess a schistosity.

With yet deeper burial into the Earth's crust, metamorphic changes continue. The general tendency, still driven by the need to minimize the energy between crystals, is for quartz and feldspar to collect together, making pale-colored bands separate from those comprising dark mafic minerals. This new segregation of minerals into bands is the characteristic of the rock called **gneiss** (Figure 6.5; see Plate 11). This is another term that undergraduate students shy away from saying out loud, though in English it simply sounds the same as "nice." The banding is properly called **gneissose banding**. It's hardly an elegant term, but it does make the distinction from the fine bedding in sedimentary rocks that geologists sometimes loosely refer to as banding.

There is no danger that the appearance of gneissose banding can be confused with the bedding of sedimentary rocks: gneisses have their coarse-grained, constituent minerals efficiently fitted together and so are dense, heavy rocks. The bands are diffusely defined and irregular, and often intricately contorted. Gneisses underlie vineyards around Planèzes, west of Rivesaltes in Languedoc-Roussillon, France, parts of the Valais hillsides, Switzerland, and the Middleburg AVA in Virginia. Under these

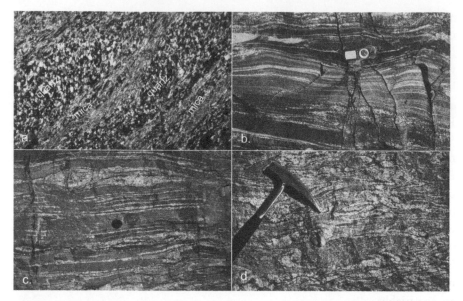

Figure 6.5 Photographs of gneiss. (a) A view down a microscope illustrating how schist grades into gneiss, as here the banding characteristic of gneiss is in its very early stages of formation. This "embryo" gneissose banding is running northeast to southwest. The granular zones consist largely of grains of quartz; the finer wispy zones consist largely of mica, representing the early stages of these two minerals partitioning into separate bands. The field of view is about 4 millimeters. (b), (c), and (d). Examples of gneiss in outcrop, with, in (d), the banding being highly folded.

intense conditions of metamorphism, even granite reorganizes its internal minerals to give a foliated look. Good examples, some with the gneissose banding highly contorted, underlie the Margaret River wine region of southwest Australia. They are spectacularly seen out on the coast, say, around the lighthouse at the southwestern-most point of Australia, Cape Leeuwin.

There is usually no visible vestige in these highly metamorphosed rocks of what the material was originally. However, by considering the overall setting of a body of gneiss in relation to adjacent rocks and by using sophisticated chemical arguments, it is sometimes possible to conclude that it was originally a sedimentary rock. In this case, it's termed a **paragneiss;** if an igneous rock: an **orthogneiss.** These are specialist terms, and many a geologist not expert in metamorphic rocks might struggle with them.

All the same, they are increasingly seen in wine writings, in northern Italy and western Austria, for example, and especially in the Muscadet region of France (Figure 6.6). The words are not always used clearly. It's incorrect to say, for example, that a vineyard is planted on gneiss and orthogneiss, as orthogneiss is a subset of gneiss. It would be like saying the birds on a feeder are both finches and goldfinches.

Figure 6.6 In vineyard soils derived from gneiss, the gneissose banding which is so clear on the clean outcrops shown in Figure 6.5 is usually much less distinct in the rock fragments. A banding is just discernible here, paralleling the platy shape of the fragments, in soils just west of Le Landreau, in the Muscadet AOC area of northwest France.

In any case, because the terms cover such a wide range of mineral and chemical compositions and resulting soils, wine labels announcing these rock types convey very little about the wine. Also, it's unclear how the geological nature of the material long before it underwent prolonged metamorphism is supposed to affect vines and wine.

What if gneiss is subjected to yet further metamorphism, even higher pressures and temperatures? There will come a point where some minerals reach their melting point; above this temperature, they will become liquid, while other minerals are still solid. In the fullness of geological time, this partially melted rock may start to be uplifted, at which point the melted portion will resolidify. Part of the rock will therefore be igneous, whereas the more refractory part that never melted will still be metamorphic. So, do we regard the exhumed rock as an igneous or a metamorphic rock? This was a subject of some dispute in the late nineteenth century, until the Finnish geologist Jakob Sederholm diplomatically suggested that it be regarded as a mixture of both, and from the Greek for mixture—*migma*—he coined the rock name **migmatite** (figure 6.7a).

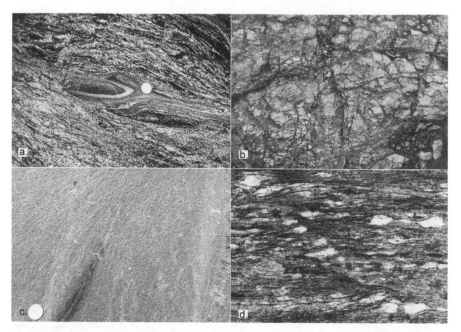

Figure 6.7 (a) Migmatite. (b) Serpentinite. Here, in the Butte de la Roche east of Nantes, in the Muscadet AOC region, the rock shows something of a snakeskin look, but the appearance can be extremely varied. The view covers about 30 centimeters. (c). Hornfels. The rock is characterized by its lack of features—the bedding of the original sedimentary rock almost completely erased by the heat-driven metamorphism. This example is at the South African winery called Hornfels. (d) The streaky, ribbon-like appearance of mylonite. The view spans about 10 centimeters.

The name "migmatite" cannot be used for just a fragment of rock; a large outcrop, say a cliff or even a mountainside, is required in order to see the assortment of patches that were once molten, typically pale colored because the pale minerals feldspar and quartz have lower melting points than the darker, unmelted parts. The arrangement of the two components is extremely variable; the whole rock often conveys a "swirly" look. Migmatite occurs immediately north and west of Kayserberg in Alsace, in the southernmost part of the Côte Rôtie in the northern Rhône Valley, and at the Corps de Loup and Coteaux de Basenon vineyards, by Tupin-et-Semons. The southern part of the Okanagan Valley, British Columbia, is underlain by migmatites; they are particularly visible in the conspicuous cliff called McIntyre Bluff, by Highway 97 south of Vaseux Lake. A wine from the Côte de Forez AOC, west of Lyon, France, is named migmatite.

Metamorphism cannot exceed migmatite. Any further temperature rise will lead to more melting, and magma will be generated, taking things into the realms of igneous rocks. Another of James Hutton's endless cycles (Chapter 1) will have started all over again. To get to the extreme stage of migmatites, our metamorphic rocks have come a long way. A wit once suggested that the long tribulations suffered by meta-morphic rocks in becoming migmatites warranted their having a special maxim. For the benefit of classically educated readers, he suggested an allusion to the motto of the British Royal Air Force *"per ardua ad astra"* (through adversity to the stars): *per ardua ad migma.*

Marble, Quartzite, and the Mysterious Serpentinite

Now we come to one of the best-known rocks of all: **marble**. But first let's note that although the rocks we will look at in this section are classed as nonfoliated (Figure. 6.1), a tectonic stress can sometimes induce some alignment features in them, and although they all consist primarily of one kind of mineral, impurities are normal. Thus, marble is composed principally of densely packed calcite but can have all kinds of appearances. The name probably conjures up a dazzling white rock, per-haps carved into gravestones, fancy hallways, statuary, and the like. That's because the great quarries of Attica, Mount Pentelicus and the Cyclades in Greece, and around Carrara in Tuscany, Italy, have produced so much marble over the millen-nia that they frame our view of the rock, and much of it is white. Similarly, picture the Taj Mahal, built of marble from Makrana, in India's Rajasthan—and a blinding white, at least from a distance.

Most of the world's marble, however, is rather darker than this. Marble is meta-morphosed limestone, so just as that rock is rarely pure, as emphasized in Chapter 5, so marble typically contains various minerals besides calcite, which gives it a mul-titude of colors and patterns. Even Carrara marble has blue-gray smudges and patches, owing to original organic carbon. Quarries for this world-famous marble dot the hillsides above the Carrara Massa wine district and over on the other side of the Alpi Apuane. Michelangelo spent many anguished times around Carrara, per-sonally seeking and extracting material that was good enough for his work. Once he narrowly missed death as a chain snapped. Now it turns out that his David, prob-ably the most famous sculpture of all, is actually a somewhat inferior marble, full of microscopic holes that are presenting challenges to those conserving the work today. Michelangelo probably used this particular marble simply because a suitably large block happened to be available, in one of the Fantiscritti quarries in Miseglia, just northeast of Carrara.

A fine pink marble occurs in some of the Minervois vineyards at Caunes, north-east of Carcassonne, France, and is still actively quarried there (it has been much used in Paris, for example in the Opera House, Arc de Triomphe, and at Versailles). However, marble is not common in the world's vineyards, and it's not a widespread rock anyway. In fact, geologically speaking, much of what is called marble is actually limestone that can be polished, which is the meaning of "marble" in the decorative stone trade. So if you encounter "marble" in vineyard regions, it may actually be limestone. You can usually tell from the other rocks that are around, especially if they contain fossils, which are common in limestone but not in marble, having been obliterated during metamorphism.

Near Swallenbach, in the Wachau, Austria, the marble in the vineyards wholly lacks fossils and is hemmed in by distinctly metamorphic rocks such as gneiss. So it's a true metamorphic marble. On the other hand, the "Comblanchien marble" that caps the skyline in parts of the Côte d'Or, France, is just one layer in a sequence of fossil-bearing sedimentary strata. It's an attractive, polishable stone, used, for example in the Basilica of St. Denis in Paris, but it's a sedimentary limestone. The Dalmation island of Brač is famous for its "marble." Away from the quarries, patches of vines are planted in depressions in the rock for wind protection (Figure 5.11), which is augmented by walls and heaps made from the white loose stones that lie all around. Although the rock is compact and polishes well, some parts are packed with delicate fossils (Figure 5.11 inset). So although it's widely called marble, geologically it's a limestone. (It's also widely said to be the stone used for the White House in Washington, D.C., but this is apocryphal. Actually, a local gray-brown sandstone was used and then painted white.)

Quartzite usually comes about through the metamorphism of quartz-rich sandstone. If the original cement was silica and metamorphism improves further the packing of the grains, then the resulting quartzite will be an exceedingly robust rock, approaching the indestructible. For example, in southern England a quartzite formed in Ordovician times, about 450 million years ago (Chapter 11), was exposed shortly after and weathered to yield quartzite pebbles. About 250 million years ago, these pebbles became incorporated in a new rock, which was buried and in turn exhumed to yield on weathering the self-same pebbles. In the geologically recent ice age, they became incorporated in glacial tills, which today are disintegrating and again delivering the pebbles, to begin yet a further incarnation. These are tough rocks! And James Hutton's vision of rocks being forever recycled (Chapter 1) is vindicated again!

Quartzite is associated with the slates of the middle Rhine Valley, including at Trechtingshausen, on the left bank north of Bingen, where a large quarry extracts durable ballast, including stone for Dutch sea defenses. It follows from this toughness that the main presence of quartzite within vineyards will be as intact stones, probably brought there by ice or rivers. Celebrated examples are the river-worn quartzite pebbles of the southern Rhône, the *galets* of parts of Chateauneuf-du-Pape.

To reemphasize one point, in view of confused usage in some wine writings, although both are composed of silica, quartzite is the *rock*, made of countless grains of the *mineral* quartz. Quartzite and quartz are not synonymous. It's a bit like a mosaic and a tile. Both are made of fired clay, but they are not the same thing. A mosaic is made of tiles; a tile is not a mosaic.

Amphibolite is typically a dense, very dark green rock, like a bundle of glittering little needles of amphibole minerals. It results from the metamorphism of mafic rocks like basalt, gabbro, and mafic tuff, and so, like those rocks, it weathers to high CEC clays like montmorillonite and may provide a good range of mineral nutrients. Amphibolite is found around Weissenkirchen in Austria's Wachau region (best seen in some of the vineyard walls) and the extreme south of the Okanagan Valley, British Columbia. In the Kutná Hora-Čáslav district just west of Prague, firmly in the beer-producing part of the Czech Republic, attempts are being made to rejuvenate an old winemaking tradition, and the metamorphic bedrock there includes amphibolite. But as with orthogneiss, the area that is trumpeting amphibolite is Muscadet, in northwest France. Amphibolite and related rocks occur sporadically there in a narrow zone stretching for about 30 kilometers southeast of Nantes.

Serpentinite is a beguiling rock. It comes about from the metamorphism of the magnesium-rich olivine in rocks like peridotite, which change to the sheet mineral serpentine. And just as with the names of rocks like quartzite, if the mineral dominates, then the rock is called serpentin*ite*. By two thousand years ago, serpentinite was already well established as an attractive decorative stone, and the Greek physician and pharmacologist Dioscorides was reporting its use as an amulet to ward off snakebite. Presumably the superstition arose because some varieties can look surprisingly like snake skin (Figure 6.7b). Actually, the rock is most variable in appearance: in addition to the snakeskin look, some examples are a very deep, lustrous, almost waxy green. One geological manual describes them as "unctuous to the touch." The strikingly dark, brooding cliffs of the southernmost point of England, The Lizard, are composed of a rich green serpentinite.

The viticultural curiosity about serpentinite is that it can actually be inimical to vines. It has long been noted that most plants can't tolerate the rock: botanists talk of "serpentinite barrens." Consequently, vines are normally not located close to the actual rock, although there are instances in France's Muscadet AOC and California's Central Coast AVA (Paso Robles, Edna Valley, and Cuesta Ridge) that involve serpentinite soils. In northern California, for example on both the Alexander Valley and Napa sides of the Mayacamas Mountains, and up in Lake County, there are barren patches where there is extremely thin soil on serpentinite bedrock. Next to them, with thicker soil, the vines can grow but may be rather stunted. If you drive up the Figuera Mountain road in Santa Barbara's Happy Canyon AVA you can see shiny, dark green serpentinite in some of the roadcuts and, looking out over the landscape, you see patches bare of vegetation underlain by this same rock. And down in the vineyards the soils with their pebbles derived from this serpentinite pose challenges for the growers. Vineyards are spread all along the southern slopes of the Troodos Mountains of Cyprus—except for the area between the Commandaria and Larnaka-Lefkosia districts, where there is extensive serpentinite.

The problem for grapevines seems to be the preponderance of magnesium in the soils, which can lead to low uptake of the nutrients calcium and potassium (Chapter 9), coupled with toxicity arising from the relatively high levels of nickel, cobalt, iron, mercury, and chromium. Growers have to apply calcium liberally

(commonly in the form of gypsum), and perhaps potassium, while carefully monitoring that the vine is absorbing the cations in an appropriate balance. Serpentinite
soils are also often unable to supply sufficient nitrogen and phosphorus to vines.

In 1965, serpentinite was officially proclaimed the "state rock" of California. It
was a curious choice, in view of the botanical difficulties, for a state where agriculture is so important and such a cornucopia of possible alternatives are available to it.
In 2010, an attempt (unsuccessful) was made to dethrone the rock, not because of
its viticultural challenges but, controversially, because it was said to contain danger-
ous asbestos. Several different kinds of minerals can naturally form as fibers and be
used as asbestos, including serpentine, but the seriously carcinogenic "blue asbestos" is an amphibole—it's not in serpentinite.

Finally, I will mention two special kinds of metamorphic rocks with names that
pop up in wines and wineries. First, there is hornfels. This rock represents the other
side of the coin, so to speak, of magma cooling to make igneous rocks. There the
heat is lost from the magma by conducting it away through the adjacent materials.
Here we are looking at the effects of this temporary heat passing through these rocks
that are near to the solidifying magma. The main result is that existing minerals may
be driven to reorganize themselves. Normally, no tectonic stress is operating, and
there is no tendency to align minerals, so the mineral grains commonly attempt
to rearrange their boundaries to form an evenly sized, roughly polygonal mosaic.
A featureless rock results, called a **hornfels**, from the word once used by miners in
Saxony (Figure 6.7c).

It follows that hornfels is a tough rock, making stony soil and displaying a resistance to weathering that may make upstanding land. Thus, it makes part of the
Kaiserstuhl hill, in the Baden wine region of Germany, and a series of narrow ridges
overlooking vineyards at the town of Pakenham, just outside the eastern limits of
Melbourne, Australia. Hornfels was quarried for roadstone in the Durban Hills,
just north of Cape Town. One winery there is actually situated in a disused quarry
(Figure 6.7c). The name of its wine? Hornfels!

The relatively quickly formed metamorphic minerals in a hornfels may have had
little chance to form a good shape, and so they commonly appear as dark, shapeless
blotches. When naming this rock, the geological pioneers did not go back to some
arcane vernacular or classical Greek but, in a masterstroke of terminological creativ-
ity, called it **spotted hornfels**. Examples occur in the Beechworth wine area, in the
High Country of northeast Victoria, Australia, and between Falset and Marçà, in
the Montsant region of northeast Spain. The black schists there, originally just like
those of nearby Priorat, have been affected by the nearby intrusion of a granite-like
rock. But the heartland of such rocks is the Andlau-Barr district of Alsace, France.
On the steep hillside overlooking the village of Andlau is the hill of Kastelberg, with
a bedrock composed of the so-called Steige schist. The name is well known to those
who revere the Grand Cru wines produced there. It's also celebrated by geologists,

Plate 1

Plate 2

Plate 3

Plate 4

Plate 5

Plate 6

a.

b.

Plate 7

Plate 8

Plate 9

Plate 10

Plate 11

Plate 12

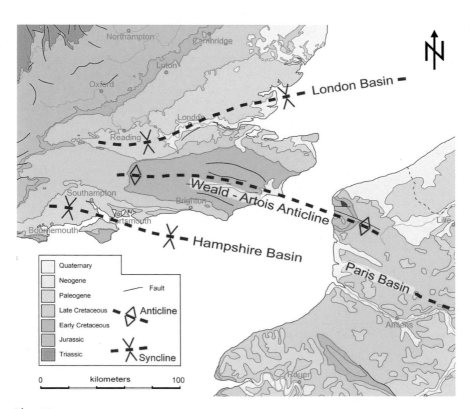

Plate 13

Plate 14

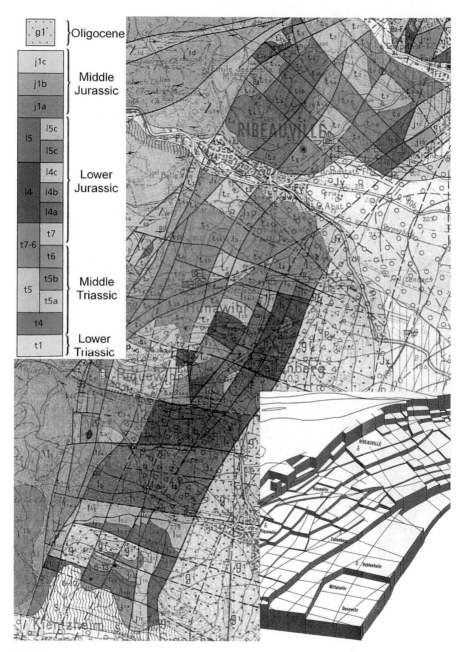

Legend (geological column):

}Oligocene — g1

Middle Jurassic — j1c, j1b, j1a

Lower Jurassic — l5 (l5c, l5c), l4 (l4c, l4b, l4a)

Middle Triassic — t7-6 (t7, t6), t5 (t5b, t5a)

Lower Triassic — t4, t1

Plate 15

Plate 16

Plate 17

Plate 18

Plate 19

Plate 20

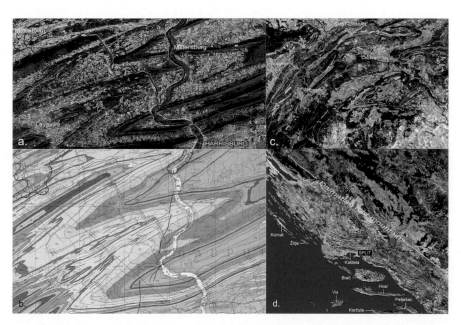

Plate 21

Plate 22

Plate 23

Geological Time Scale (International Chronostratigraphic Chart)

Table 1

Eon	Era	Period	Epoch	Age	Numerical age
Phanerozoic	Cenozoic	Quaternary	Holocene		present / 0.0117
Phanerozoic	Cenozoic	Quaternary	Pleistocene	Upper	0.126
Phanerozoic	Cenozoic	Quaternary	Pleistocene	Middle	0.781
Phanerozoic	Cenozoic	Quaternary	Pleistocene	Calabrian	1.80
Phanerozoic	Cenozoic	Quaternary	Pleistocene	Gelasian	2.58
Phanerozoic	Cenozoic	Neogene	Pliocene	Piacenzian	3.600
Phanerozoic	Cenozoic	Neogene	Pliocene	Zanclean	5.333
Phanerozoic	Cenozoic	Neogene	Miocene	Messinian	7.246
Phanerozoic	Cenozoic	Neogene	Miocene	Tortonian	11.63
Phanerozoic	Cenozoic	Neogene	Miocene	Serravallian	13.82
Phanerozoic	Cenozoic	Neogene	Miocene	Langhian	15.97
Phanerozoic	Cenozoic	Neogene	Miocene	Burdigalian	20.44
Phanerozoic	Cenozoic	Neogene	Miocene	Aquitanian	23.03
Phanerozoic	Cenozoic	Paleogene	Oligocene	Chattian	28.1
Phanerozoic	Cenozoic	Paleogene	Oligocene	Rupelian	33.9
Phanerozoic	Cenozoic	Paleogene	Eocene	Priabonian	37.8
Phanerozoic	Cenozoic	Paleogene	Eocene	Bartonian	41.2
Phanerozoic	Cenozoic	Paleogene	Eocene	Lutetian	47.8
Phanerozoic	Cenozoic	Paleogene	Eocene	Ypresian	56.0
Phanerozoic	Cenozoic	Paleogene	Paleocene	Thanetian	59.2
Phanerozoic	Cenozoic	Paleogene	Paleocene	Selandian	61.6
Phanerozoic	Cenozoic	Paleogene	Paleocene	Danian	66.0
Phanerozoic	Mesozoic	Cretaceous	Upper	Maastrichtian	72.1
Phanerozoic	Mesozoic	Cretaceous	Upper	Campanian	83.6
Phanerozoic	Mesozoic	Cretaceous	Upper	Santonian	86.3
Phanerozoic	Mesozoic	Cretaceous	Upper	Coniacian	89.9
Phanerozoic	Mesozoic	Cretaceous	Upper	Turonian	93.9
Phanerozoic	Mesozoic	Cretaceous	Upper	Cenomanian	100.5
Phanerozoic	Mesozoic	Cretaceous	Lower	Albian	~113.0
Phanerozoic	Mesozoic	Cretaceous	Lower	Aptian	~125.0
Phanerozoic	Mesozoic	Cretaceous	Lower	Barremian	~129.4
Phanerozoic	Mesozoic	Cretaceous	Lower	Hauterivian	~132.9
Phanerozoic	Mesozoic	Cretaceous	Lower	Valanginian	~139.8
Phanerozoic	Mesozoic	Cretaceous	Lower	Berriasian	~145.0

Table 2

Eon	Era	Period	Epoch	Age	Numerical age
Phanerozoic	Mesozoic	Jurassic	Upper	Tithonian	152.1
Phanerozoic	Mesozoic	Jurassic	Upper	Kimmeridgian	157.3
Phanerozoic	Mesozoic	Jurassic	Upper	Oxfordian	163.5
Phanerozoic	Mesozoic	Jurassic	Middle	Callovian	166.1
Phanerozoic	Mesozoic	Jurassic	Middle	Bathonian	168.3
Phanerozoic	Mesozoic	Jurassic	Middle	Bajocian	170.3
Phanerozoic	Mesozoic	Jurassic	Middle	Aalenian	174.1
Phanerozoic	Mesozoic	Jurassic	Lower	Toarcian	182.7
Phanerozoic	Mesozoic	Jurassic	Lower	Pliensbachian	190.8
Phanerozoic	Mesozoic	Jurassic	Lower	Sinemurian	199.3
Phanerozoic	Mesozoic	Jurassic	Lower	Hettangian	201.3
Phanerozoic	Mesozoic	Triassic	Upper	Rhaetian	~208.5
Phanerozoic	Mesozoic	Triassic	Upper	Norian	~227
Phanerozoic	Mesozoic	Triassic	Upper	Carnian	~237
Phanerozoic	Mesozoic	Triassic	Middle	Ladinian	~242
Phanerozoic	Mesozoic	Triassic	Middle	Anisian	247.2
Phanerozoic	Mesozoic	Triassic	Lower	Olenekian	251.2
Phanerozoic	Mesozoic	Triassic	Lower	Induan	252.17
Phanerozoic	Paleozoic	Permian	Lopingian	Changhsingian	254.14
Phanerozoic	Paleozoic	Permian	Lopingian	Wuchiapingian	259.8
Phanerozoic	Paleozoic	Permian	Guadalupian	Capitanian	265.1
Phanerozoic	Paleozoic	Permian	Guadalupian	Wordian	268.8
Phanerozoic	Paleozoic	Permian	Guadalupian	Roadian	272.3
Phanerozoic	Paleozoic	Permian	Cisuralian	Kungurian	283.5
Phanerozoic	Paleozoic	Permian	Cisuralian	Artinskian	290.1
Phanerozoic	Paleozoic	Permian	Cisuralian	Sakmarian	295.0
Phanerozoic	Paleozoic	Permian	Cisuralian	Asselian	298.9
Phanerozoic	Paleozoic	Carboniferous (Pennsylvanian)	Upper	Gzhelian	303.7
Phanerozoic	Paleozoic	Carboniferous (Pennsylvanian)	Upper	Kasimovian	307.0
Phanerozoic	Paleozoic	Carboniferous (Pennsylvanian)	Middle	Moscovian	315.2
Phanerozoic	Paleozoic	Carboniferous (Pennsylvanian)	Lower	Bashkirian	323.2
Phanerozoic	Paleozoic	Carboniferous (Mississippian)	Upper	Serpukhovian	330.9
Phanerozoic	Paleozoic	Carboniferous (Mississippian)	Middle	Visean	346.7
Phanerozoic	Paleozoic	Carboniferous (Mississippian)	Lower	Tournaisian	358.9

Table 3

Eon	Era	Period	Epoch	Age	Numerical age
Phanerozoic	Paleozoic	Devonian	Upper	Famennian	372.2
Phanerozoic	Paleozoic	Devonian	Upper	Frasnian	382.7
Phanerozoic	Paleozoic	Devonian	Middle	Givetian	387.7
Phanerozoic	Paleozoic	Devonian	Middle	Eifelian	393.3
Phanerozoic	Paleozoic	Devonian	Lower	Emsian	407.6
Phanerozoic	Paleozoic	Devonian	Lower	Pragian	410.8
Phanerozoic	Paleozoic	Devonian	Lower	Lochkovian	419.2
Phanerozoic	Paleozoic	Silurian	Pridoli		423.0
Phanerozoic	Paleozoic	Silurian	Ludlow	Ludfordian	425.6
Phanerozoic	Paleozoic	Silurian	Ludlow	Gorstian	427.4
Phanerozoic	Paleozoic	Silurian	Wenlock	Homerian	430.5
Phanerozoic	Paleozoic	Silurian	Wenlock	Sheinwoodian	433.4
Phanerozoic	Paleozoic	Silurian	Llandovery	Telychian	438.5
Phanerozoic	Paleozoic	Silurian	Llandovery	Aeronian	440.8
Phanerozoic	Paleozoic	Silurian	Llandovery	Rhuddanian	443.8
Phanerozoic	Paleozoic	Ordovician	Upper	Hirnantian	445.2
Phanerozoic	Paleozoic	Ordovician	Upper	Katian	453.0
Phanerozoic	Paleozoic	Ordovician	Upper	Sandbian	458.4
Phanerozoic	Paleozoic	Ordovician	Middle	Darriwilian	467.3
Phanerozoic	Paleozoic	Ordovician	Middle	Dapingian	470.0
Phanerozoic	Paleozoic	Ordovician	Lower	Floian	477.7
Phanerozoic	Paleozoic	Ordovician	Lower	Tremadocian	
Phanerozoic	Paleozoic	Cambrian	Furongian	Stage 10	~489.5
Phanerozoic	Paleozoic	Cambrian	Furongian	Jiangshanian	~494
Phanerozoic	Paleozoic	Cambrian	Furongian	Paibian	~497
Phanerozoic	Paleozoic	Cambrian	Series 3	Guzhangian	~500.5
Phanerozoic	Paleozoic	Cambrian	Series 3	Drumian	~504.5
Phanerozoic	Paleozoic	Cambrian	Series 3	Stage 5	~509
Phanerozoic	Paleozoic	Cambrian	Series 2	Stage 4	~514
Phanerozoic	Paleozoic	Cambrian	Series 2	Stage 3	~521
Phanerozoic	Paleozoic	Cambrian	Terreneuvian	Stage 2	~529
Phanerozoic	Paleozoic	Cambrian	Terreneuvian	Fortunian	541.0

Plate 24

for this is the very rock where the nature of hornfels, spotted and otherwise, was elucidated.

The next chapter changes subjects, discussing geological faults and explains how large-scale examples occur as zones that can contain highly deformed rocks. At conditions deep in the Earth's crust, the material inside this narrow zone can become extremely drawn out, with each mineral grain becoming tremendously extended such that a metamorphic rock develops, with a striking ribbon appearance. From the idea that the rock was being milled, the name **mylonite** was coined. The rock is dense, with a glassy, "streaked out" look (Figure 6.7d). Being restricted to certain fault zones, mylonites are uncommon in vineyards. However, they occur in the old metamorphic rocks of the Virginia Piedmont, in the central Virginia wine region. One Virginia winery features a fine picture of a mylonite on its eponymous wine label.

It may be a somewhat obscure rock, but it seems that mylonite's name is popular: it's the title of a student magazine in California, and you can buy Mylonite sports shoes, not to mention Mylonite ladies dresses. The English geologist who coined the word, Charles Lapworth (1842–1920), wrote in his diary that when he was struggling to interpret how the rock formed, for several nights he suffered nightmares. He would wake perspiring, having dreamed that he was "bodily caught up" in what he called "the great earth engine," as though it was he who was being stretched out and milled. Trying to understand metamorphic rocks can be an intense experience!

Further Reading

There are a number of populist booklets on metamorphic rocks with titles telling us that "the earth rocks" and adding words like squashing, crunching, squeezing, and crushing. In contrast, weighty treatises have been written, announcing in their titles such things as petrogenesis, geodynamics, and phase equilibria. Perhaps this chasm between the two genres is best bridged by the relevant parts of books that cover all the rock types, and even all the common rocks and minerals. Suggestions include the following:

Farndon, John. *The Practical Encyclopedia of Rocks and Minerals: How to Find, Identify, Collect and Preserve the World's Best Specimens*. With over 1000 Photographs and Artworks. London: Hermes House, 2015.

Grice, Joel. *Beginner's Guide to Minerals and Rocks*. Markham, Ontario: Fitzhenry and Whiteside, 2010.
Grice, a curator at the Canadian Museum of Nature in Ottawa, has aimed his book primarily at collectors, but it has an unrivalled selection of color photographs.

Hollocher, Kurt. *A Pictorial Guide to Metamorphic Rocks in the Field*. Boca Raton, FL: CRC Press, 2014.
Hollocher's work has a lot of technical material and terminology well beyond that introduced in this chapter but it also has many good photographs of metamorphic rocks.

The Rocks Change Shape

Folds, Faults, and Joints

Even Rocks Can Change Their Shape

You may have looked at some rocky cliff and noticed sedimentary strata bent into huge curves, the shapes that geologists call folds. You may even have heard of terms like anticline and syncline. Almost certainly you will have heard of geological faults: the San Andreas Fault in California must be one of the best-known geological features there is. They are all examples of what geologists call geological structures. They can affect vineyards, and the names of examples appear around the world on wine labels. So how do these structures come about, and what decides whether rocks make folds or faults?

We introduced the concept of tectonic stresses in the previous chapter. We learned that because they act in a particular direction they can induce foliations within metamorphic rocks, but of relevance here is that they can also cause rocks to change their overall shape. That is, the rocks **deform**, which gives rise to various **geological structures**. Any solid matter (unlike a liquid) that feels stresses, of whatever origin, will resist them *up to a point* before it starts to change shape. That point is what defines the **strength** of the material. The same principles apply when stresses are applied to a sediment or a soil, though rocks, with their constituent minerals firmly bonded together, resist much greater levels of stress before they deform. As one wag put it, the difference between a rock and a soil is that when you kick them a rock hurts your foot . . .

So, focusing in on rocks, we see two ways in which they can deform: by flow and by fracture. Looking ahead to where this is going to lead, it's flow that gives rise to folds, and faults result from fracture. A good analogy for the **flow** of rocks is glacial ice. The ice is solid to us, but given time, it can flow, to give the "river of ice" that is a glacier. If you leave a ball of silicone putty on a table top, after a few days it will have flowed, while still being a solid, to make a pool. Window glass in very old buildings has flowed such that it is thicker at the bottom. It's a matter of available time: rocks take that much longer to flow. Over millions of years, the overall shape of a rock can change but it is still intact, in a coherent mass. In contrast, **fracture** involves loss of such cohesion; the rock breaks into two or more separated parts. So the question of whether rocks make folds or faults is really a matter of whether the rocks flow or fracture, and there are several things that decide this, as shown in the accompanying box.

Rocks Can Fracture and Rocks Can Flow—What Decides?

The pressure that acts on rocks residing at some depth within the Earth—burial pressure—tends to induce flow. This is because when a material fractures, it increases in bulk volume very slightly as the pieces come apart. Burial pressure works to inhibit this expansion and so promotes deformation by flow instead.

Working hand in hand with the greater burial pressure at depth below the ground is increased temperature, which also promotes flow. Think of breaking a bar of chocolate. While it will readily fracture if the bar is fresh out of the refrigerator, if it has been in your pocket for some time it will tend to flow rather than break. It's the same with trying to spread butter that has been refrigerated as opposed to butter having sat at room temperature. In neither instance has the material melted, in which case it obviously will flow as a liquid, but temperature influences the way the solid substance deforms. So because both burial pressure and temperature increase with depth down into the Earth, in general rocks that are near the land surface tend to fracture, whereas exactly the same rock will flow at depth.

Another factor is the time taken for the deformation to take place. As we have seen, flow requires time, so slow rates tend to promote flow, and rapid deformation gives fracture. Imagine grasping a ball of a child's play putty with both your hands. If you yank the material apart, the chances are you will end up with two pieces of putty, one in each hand—that is, the material will have fractured. But if you pull the material apart very, very slowly, you can probably get it to stretch out thinly yet still be in one intact piece; it has accomplished an enormous shape change by flowing. In both cases, it is exactly the same material, at the same temperature and the same "burial pressure" (atmospheric in this case), but the contrasting deformation has come about purely through the differing rates.

So, once the strength of a rock is exceeded it either fractures or it flows. In some circumstances, the resulting structures in nature may have involved both flow and fracture: for example, the rock may have started to flow, but then because something changed, fracture took over instead. Conversely, many faults are accompanied by a degree of folding. Again a glacier provides an analogy: the river of ice can be flowing but with arrays of the notorious crevasses, which are fractures in the ice. Also, a later structure can be superimposed on an earlier one, being of different nature because the conditions have changed. You can gaze at the smoothly curving lines of folded strata in a cliff, clearly implying flow of the rocks while at some depth, but you can now bash the cliff with a hammer and shards will break off. Your instantaneous application of stress at surface conditions has made the same rock deform very differently!

Bending and Bowing: The Folding of Rocks

Rocks flow to produce a variety of geological structures, but of greatest relevance here is the effect on a pile of layered sedimentary rocks. Providing the tectonic stress is directed parallel or at a low angle to the stratification, the sedimentary beds will flow into a series of undulations called **folds**. The process is known as **folding**. In principle, any surfaces in rocks can be folded, not just sedimentary layers, but the deforming stresses have to work roughly parallel to them. If you arrange to squeeze a pile of carpets and do it hard enough—that is, exceed the strength of the pile—then they will fold, provided you are compressing along the length of the carpets. Self-evidently, you will never create such undulations if you are squeezing across the pile.

In nature, we are usually unable to see much of the train of folds developed in the layers but just a segment, perhaps an individual fold or two (Figure 7.1; see

Figure 7.1 Folded strata. A syncline and anticline in the cliffs at St. Anne's Head, Pembrokeshire, Wales.

Plate 12). In this case, we refer to the upwarp or arch-like part as an **anticline** and the downwarped trough-like part as a **syncline**. In the Columbia Gorge AVA, the Syncline Winery took its name from the U-shaped structure in the bedrock on which it is situated.

In highly deformed parts of the Earth's crust, such as the interior areas of some mountain chains, the folds can be highly convoluted and in any orientation. In Italy's Valle d'Aosta DOCG, there are stunning contortions of schists and gneisses, probably best seen at Donnas on the Roman road that once connected Rome with the Rhône Valley in France. (The gneisses were often used to build the stone pergolas for the traditional system of training vines, not to mention the *pietre*, sizzling-hot stone slabs for cooking the local pork.) On the Roman road, hewn into the rock by slaves and still showing the wheel ruts made by passing carriages, you can see the schistosity and the gneissose banding twisted into wildly tortuous swirls. However, in most cases relevant to vineyards, the train of folds will average out as a roughly horizontal sheet, just like the undulating pile of carpets, and we can think simply of anticlines and synclines as reasonably symmetrical upwarps and downwarps.

Folds can be very gentle, with the strata on their flanks not deviating greatly from the horizontal. Even so, on a regional scale, they can have important effects on the distribution of rocks at the land surface and hence on viticulture. The arrangement of the sedimentary strata in the vine-growing areas of southeast England and northwest France (Figure 7.2; see Plate 13) is an example. (Of course, the English Channel developed in relatively recent geological times, and cut right across these already formed structures.)

Another repercussion of folding is its generating hillslopes, influencing both their gradient and their aspect, that is, to what extent they face toward the sun. In the Columbia Valley AVA, the gently inclined south-facing slopes of anticlines receive 3 to 5% more solar energy per unit area than the relatively flat centers of the synclinal basins. Cold air drains from the flanks of those anticlines to pool in the adjacent synclinal troughs; hence, that is where vineyards with the greatest risk of frost and freeze damage are located. Some of the warmest vineyards in the Columbia River region are located close to the crests of anticlines that lie above the nocturnal cold air pool.

Many of the Columbia Valley AVAs are defined largely by a topography that is a consequence of fold structures. The Rattlesnake Hills, Red Mountain, and Horse Heaven Hills AVAs, for example, are all situated on the south-facing flanks of anticlines, and the Walla Walla Valley and Yakima Valley AVAs coincide with synclinal valleys. The boundaries of the Snipes Mountain AVA encircle an anticlinal ridge.

However, an anticline does not necessarily correspond to a dome of the land surface, nor does a syncline necessarily correspond to a land trough; the form of the ground depends on the relative toughness of the strata and not the shape of the folds. A classic example occurs in the North Hampshire Downs of southern

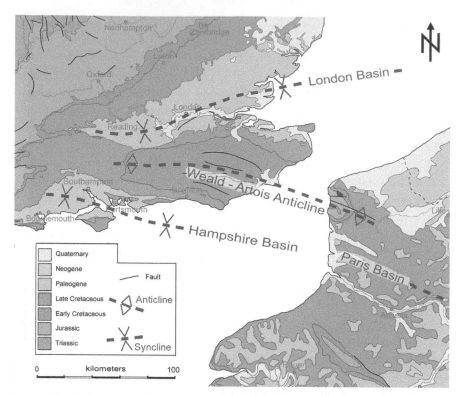

Figure 7.2 Geologic map of southeastern England and the Boulonnais, northwest France, showing the influence of regional-scale folds on the distribution of strata. Based on a map at https://commons.wikimedia.org/wiki/File:Geologic_map_SE_England_%26_Channel_EN.svg.

England, in the Vale of Kingsclere (which continues westward as Pewsey Vale, a name also familiar to Barossa wine lovers). Figure 7.3 represents the first documentation of how folded strata and landscape relate. The valley has been eroded in the softest rocks, which happen to be in the core of an anticline. Thus, the bedrock strata on the north side of the valley, the northern part of the anticline, slope away northward, but the southern flank gives south-facing, well-drained slopes, just the kind of situation utilized by English vineyards. The area in this example was made famous in the book *Watership Down*, but it is now home to a flourishing sparkling wine industry as well as rabbits.

A further consequence of folding is that beds at differing levels in the sedimentary pile are brought to the same land elevation. In an anticline, what were the lowest and hence first deposited strata are raised in the central portion of the fold. Thus, if erosion carves a roughly horizontal land surface through the folded pile, the anticlines will have the oldest strata in their central parts. Synclines will depress the higher, younger strata in their centers. This is why in the Côte d'Or (Figure 7.4), the

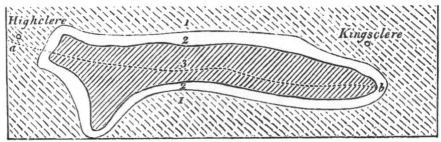

Valley of Kingsclere.

a, b, Anticlinal line marking the opposite dip of the strata on each side of it.

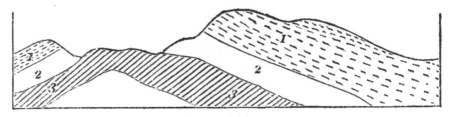

Section across the Valley of Kingsclere from north to south.

1, Chalk with flints.   2, Lower chalk without flints.
3, Upper green-sand, or firestone, containing beds of chert.

Figure 7.3 Geologic map of the Vale of Kingsclere, in Hampshire, between Newbury and Basingstoke, England, the first documentation of two important geological principles: (1) an anticline, though an upwarp in the bedrock, does not necessarily form higher ground, and here it forms a valley; (2) the rocks become symmetrically younger outward from the core of the fold. From an 1829 paper by William Buckland tellingly entitled: "On the Formation of the Valley of Kingsclere and Other Valleys by the Elevation of the Strata That enclose Them," *Transactions of the Geological Society of London,* 2nd ser., **2** (1829), 119–130.

Côte de Beaune tends to be dominated by younger, once higher, strata flexed down in the so-called Volnay syncline, whereas to the north, for example, around Nuits St. Georges, older formations of limestone are bowed up in the Gevrey anticline.

Very gentle folds aligned northeast to southwest control the distribution of strata on the right bank of the Gironde estuary in France. If you follow Route D669 along the north bank of the Dordogne River from Bourg toward Blaye, a direction that happens to be roughly southeast to northwest, you cross progressively younger layers on the southern flank of the so-called Bourg-Listrac anticline to reach the fold's core at Blaye. So although the ground keeps more or less the same elevation, you begin on relatively young bedrock, limestones of Oligocene age, whereas just below the Citadel in Blaye, the fossil-rich sandstones and limestones are of Middle Eocene age, the oldest visible bedrock in the district. Figure 7.5 shows the general arrangement and how the folds continue in the bedrock below the surface.

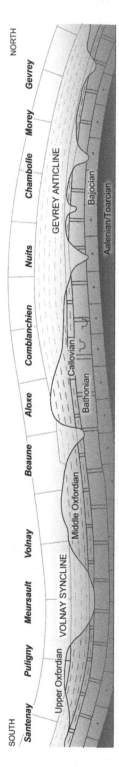

Figure 7.4 A sketch profile along the Côte d'Or, France, showing how the Volnay syncline and Gevrey anticline govern the distribution of bedrock strata.

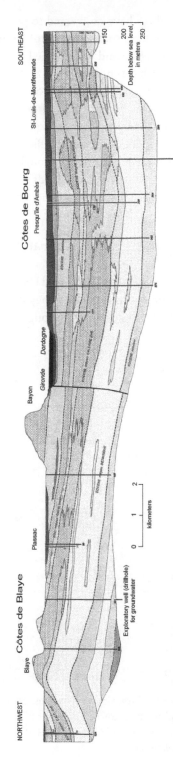

Figure 7.5 Geologic profile along the right bank of the Gironde estuary showing how regional open folds control the distribution of strata at the land surface, for example, in the Côtes de Bourg and de Blaye, and in the bedrock below. The strata consist of shales, sandstones, marls, limestones, and so on, mainly of Oligocene and Eocene age. (Figure 11.4 explains these geological time names.) The vertical lines are wells (drill-holes), which provided much of the underground information; the dashed lines are related to groundwater levels. Adapted from a BRGM report on the *Balades hydrogéologiques en Aquitaine* (http://sigesaqi.brgm.fr/IMG/pdf/balade_hydrogeologique_-_secteur_de_blaye_-_introduction.pdf). © BRGM, Longitudinal section along the Gironde estuary—Authorisation R16/21.

In some situations where tectonic plates have converged to produce mountain ranges, long linear zones of intensely folded rocks result: **fold mountains**. The Appalachians are an example, and Figures 8.7a and b show the landform pattern in part of Pennsylvania. Some wineries there have chosen to utilize the valley bottoms, frost prone but sheltered by the sharp quartzite ridges that loop around them; others have chosen to use the narrow hilltops, including some of the highest American wineries east of the Rockies. The contorted bedrock that forms the mountains of Dalmatian Croatia produces the series of rugged limestone cliffs that hug the coastal areas and that continue out to sea, under water. In places, the tough limestone ridges appear above sea level, giving numerous long, narrow islands and peninsulas, all with important wine-making traditions (see Chapter 8 and Figure 8.7d).

Cracking and Moving: Geological Faults

Faults are fractures in rock along which either side has moved (Figure 7.6; see Plate 14). (Thus, they differ from joints, to be discussed later.) They can be of any size, ranging from microscopic up to hundreds of kilometers; as long as the crack shows

Figure 7.6 Example of faults in rocks. The sandstone layers have been abruptly displaced along fractures, effectively "downstepping" the beds toward the right of the photograph. The view is about 5 meters across. Location is alongside U.S. Route 191 north of Moab, Utah, just before the entrance to Arches National Park.

relative displacement, it's still a fault. They're three dimensional, even though we tend to think of just the lines they make at the Earth's surface. Indeed, the word "fault" has entered popular parlance in this way, such as in "a political fault line." Rarely is a fault a single fracture: fractures are usually narrow zones of breakage, and larger structures commonly have subsidiary faults splaying off them.

Some believe that wines from vineyards with faults running through them have a special "dynamism," a "verve" and a "focused energy." But there's nothing mystical about faults; they possess no "vortexes or mysterious forces." Faults can, however, have a number of repercussions for viticulture. First, a fault juxtaposing different rocks can change the pattern of groundwater flow in the bedrock, and the fault zone itself may have an effect. Some vine growers maintain that a fault running through their vineyard reduces groundwater flow, through acting as a barrier, whereas others claim that a fault improves groundwater flow, citing lines of springs developed along the fault line. Both claims can be true: it differs between faults.

Second, the fault zone itself has an effect on the land surface. Almost invariably, the rocks within a fault zone are weak because of the internal breakage, and so they are commonly picked out by erosion. For example, in the Roussillon region of France, the River Têt, which forms the boundary between the Côtes de Roussillon and Côtes de Roussillon-Villages appellations, is eroding along the North Pyrenean Fault and to the west it is eroding into the Pyrenees, past Prades, along a subsidiary fault that splays off to the southwest.

Then there are the topographic effects that result from the rocks differing on either side of a fault. For example, the northeastern edge of the northern Barossa Valley, Australia, around Stockwell (Figure 1.5b), is defined by a fault that brings the softer, young sediments of the valley against the much older, harder rocks that make the hills to the east, forming a **fault escarpment** *or* **fault scarp**. There are examples in the coarse gravels along the Huangarua River in Wairarapa, New Zealand, and one famous winery nearby is known simply as "Escarpment."

The scarp itself, as a narrow zone of relatively steep slopes separating two faulted blocks, can be important. In Burgundy, along the roughly north–south zone of interweaving breaks known collectively as the Saône Fault, the land to the west was uplifted, to give the Hautes Côtes and higher land stepping up westward. To the east, the downdropped plains are now occupied by the Saône River and its deposits, on which grow vines traditionally producing Bourgogne ordinaire (now Coteaux Bourguignons). Caught in between is the vinous honeypot of the Côte d'Or.

Eastward across the Rhône Valley from here is another fault that is something of a mirror image, bringing up to its east the Jura and Savoie regions. In other words, the river flows in a downdropped block between these two faults, in a structure known as a **rift valley** or **graben**. In Burgundy, the structure is called the Saône Graben, and further afield it is called the Bresse Graben. Large-scale examples (such as the East African Rift) are incipient divergent plate boundaries, the "Continental rift zone" of Figure 1.8b, though the Bresse Graben never grew into a plate boundary. Not all places called graben are rift valleys, though; the word is also simply the German for a ditch or trench. Thus, the famous vine-laden Spitzer Graben that strikes eastward behind the Wachau village of Spitz on the Danube is a pronounced valley, but it isn't a geological structure; it isn't formed by faults.

The upfaulted block that complements a graben is known as a **horst**. Part of the block to the west of the Saône Graben, southwest of the Chalonnais, is called the Horst de Mont-St. Vincent. Northward from Burgundy, the graben continues as the Rhine Graben, with the Vosges Horst to the west and the Black Forest Horst to the east. The north–south strip of classic vineyards on the western scarp of the Rhine Graben, in Alsace, contains an astonishing variety of bedrock because of the faulting (Figure 7.7; see Plate 15). Basically, the faults have brought younger alluvial sediments next to the very much older igneous and metamorphic rocks of the Vosges Horst, with all kinds of sedimentary rocks caught up in between the subsidiary faults.

The setting where such intricate changes are most pronounced is with fault zones that are overall at a very low angle, nearly horizontal. These are **thrust faults**. A simplistic analogy is an intact deck of cards resting on a table that is given a nudge. The cards will smear out, overlapping one another. In nature, the "push" is ultimately to do with a tectonic plate driving below another one at a low angle, though some faults splay off upward at high angles, finely slicing up the rock formations. A number of wine districts around the world are located in or next to such zones; examples are St-Chinian, in the Languedoc, France (Figure 7.8), Liébana in Cantabria, northern Spain, and the Sierra Madre Oriental, in east-central Mexico just north of the expanding wine district of Querétaro Province.

Around the village of Durban in Corbières, southern France, soils can change from gray sands to detritus of dark schist to salmon-pink limestone soils in just a few meters, although only in places can the faults responsible for these sudden changes in the bedrock be seen. Thrust faults usually follow the weakest layers, which hereabouts often means those rich in soft gypsum. Roadcuts along the east of the D77 south of Durban show spectacularly convoluted gypsum layers, deformed during the faulting. However, just east of the D106 road south of Coustouge, to the northwest of Durban, a fault climbs through coarse gravels, actually fracturing and displacing some of the pebbles.

A variation on thrust faulting is classically developed in the Jura area of northeast France (Figure 7.9). In 1907, August Buxtorf, a Swiss geologist at the University of Basel, was looking at some new railroad tunnels and came up with an idea that was to remain controversial for decades. The low-angle faults that he saw, developed along layers of gypsum and salt, only seemed to be displacing and deforming the rocks above them, not those below. Nowhere could he see the faults diving down, as thrust faults do, to link in with the deep stresses of the Earth. So he hypothesized that somewhere these faults must curve upward to emerge at the land surface and that everything above them had simply glided along their slight downward slope

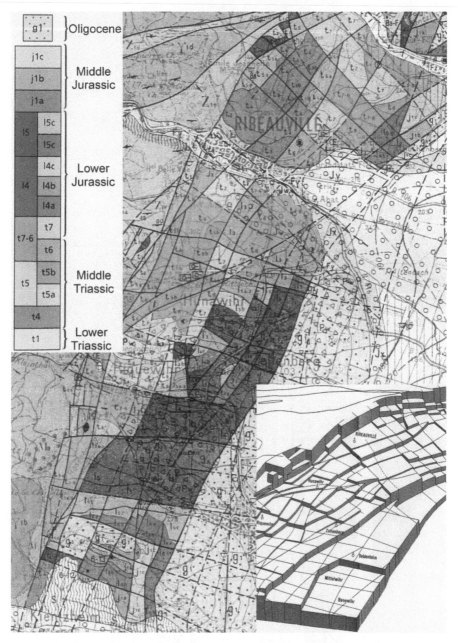

Figure 7.7 Intricate fault system apparent in the geologic map of Ribeauvillé, Alsace, France. Extract of geologic map no. 342 (Artolsheim-Colmar) at 1:50,000 (© BRGM, Authorisation R16/21). Inset is adapted from a diagram in a geologic field guide to the region, at http://www.lithotheque.site.ac-strasbourg.fr/pres-de-chez-vous/centre_alsace/florimont/florimont-fiche-professeur.

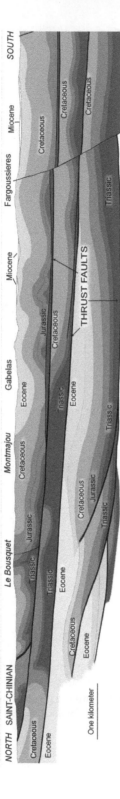

Figure 7.8 Schematic geologic cross section south of St-Chinian, Languedoc. Sedimentary strata are displaced along very low-angle thrust faults, shown as bold black lines, with some steeper upward splays. The thrusts have displaced younger beds up and over slices of older strata. Toward the south, a steeply inclined, later fault displaces downward all the beds and thrust faults. Based on information on the BRGM 1:50,000 1014 St-Chinian and 1039 Bézier maps. Compare with Figure 7.9.

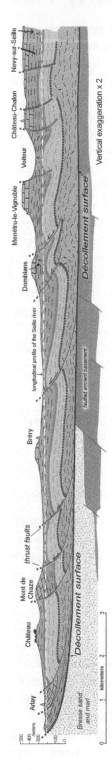

Figure 7.9 Geologic cross section west to east across the Jura vineyard region showing the master décollement developed in weak gypsum layers, with upward splaying faults that displaced the overlying strata. Adapted from a figure in Wink Lorch, *Jura Wine*, 2014, by kind permission of Michel Campy and Wink Lorch.

due to gravity. Buxtorf interpreted the process as an "ungluing" of the rocks and called it by the equivalent French word **décollement**. The mechanism is now well documented in many parts of the world, and décollement is the established term in English for this kind of fault, though some geologists prefer **detachment**. The distribution of vineyards in the Jura district is closely related to the results of this décollement.

Another important type of fault, variously called a **strike-slip, wrench, tear**, or **transverse** fault, refers to displacement across the two sides that is not "up and down" but sideways, roughly in the horizontal, parallel to the land surface. These faults are infrequent in distribution but have enormous significance where they do occur. The largest examples are transform plate boundaries (Figure 1.8b), exceeding 1000 kilometers in length and showing displacements of tens, even many hundreds, of kilometers. Not surprisingly then, strike-slip faults are typically complex fault zones, with intricate displacements within the zone and plenty of subsidiary splay faults.

Moreover, strike-slip faults go down deep into the Earth's crust, creating a permanent zone of weakness such that if stresses subsequently build up in the Earth it may well be these fault zones that give first. If this happens, the accumulated energy is abruptly released—as sound, as heat, and as the physical waves which on reaching the Earth's surface we call **earthquakes**. Actually, the greater part of the Earth's earthquake energy is dispersed from those places where a tectonic plate is grinding *below* another one (in what is effectively a mega-thrust fault). Otherwise the most damaging on-land earthquakes are associated with repeated displacements on existing strike-slip faults. Damage from earthquakes is, of course, yet another way faults can affect vineyards.

"The Earthquake Had a Magnitude of 6." What Does That Mean?

Earthquake studies provide another geological field with technical terms that have made their way into everyday language, such as those beginning with "seism . . .," derived from an old Greek word for shaking. **Seismology** *is the scientific study of* **seismic waves** *(pronounced size-mik) and so on, but we now read of seismic changes in politics and business and in technology; and apparently we're living at "a seismic time in the history of marijuana and culture." Another word with a formidably technical origin but now in popular use is the* **Richter scale***, as in suggestions of a Richter scale of racism, power outages, and sex. Properly, however, it's a way of numerically comparing the sizes of earthquakes.*

Seismologists were for a long time challenged to explain how to quantify earthquakes, as the obvious ways involving the amount of damage depend on factors like the kinds of places being affected and on their distance from the earthquake's origin, as much as the earthquake itself. (Incidentally, that initiating point within the Earth is technically called the earthquake's **focus***; in science, the well-known word*

epicenter *refers to the place that happens to be above it, at the ground surface.) In 1935, Charles Richter and co-workers finally devised a mathematical way of using the size of the wiggles in the ink trace of a seismograph, suitably calibrated and with adjustments for distance, to indicate the amplitude of the seismic waves—in other words to quantify the undulations of the shaking ground. Working at Caltech, they found that the system worked reasonably well for California, after which it caught on more widely. Richter adopted a logarithmic method for the numbers (i.e, in increments of 10, just like pH, see Chapter 10), in order to compress the huge variations in seismic amplitude into a convenient scale. This means that each whole-number division on the Richter scale differs by 10. Thus, an earthquake that registers 6.0 on the scale has a shaking amplitude that is 10 times greater than one of 5.0 and 100 times greater than an event measuring 4.0.*

The much-loved Richter scale is now obsolete. Earthquakes these days are expressed by something called the moment magnitude, or M_W (the w subscript stands mathematically for the work done). This measurement avoids most of the assumptions and restrictions of the Richter method and focuses on the energy released during the earthquake rather than on the amount of ground shaking. It turns out that the actual energy released by different earthquakes, the vigor of the shaking as it were, differs by a much larger amount than the shaking amplitude—around 31 times more. So although this new scale uses numbers that in most situations are approximately equivalent to the Richter values, it's important to realize that each whole-number value differs from its neighbor by a factor not of 10 but of 31. It makes a big difference. Journalists reporting an earthquake don't always treat these magnitude numbers with sufficient care: vaguely saying that an earthquake was something like a 6 or a 7 can be misleading. For example, an earthquake measuring 8.7 on the M_W scale is about 23,000 times more energetic—stronger—than a 5.8.

Even so, that the effects of earthquakes depend on considerations other than sheer physical magnitude is well illustrated by the 2010–2011 events in the Canterbury area of New Zealand. The main shock in 2010 had an M_W magnitude of 7.0 but appears to have led to only two injured persons and one fatality—and indirectly at that. Yet, an aftershock six months later with the lesser magnitude of 6.3 injured several thousand people and led to 185 fatalities. In even greater contrast, the awful Haiti earthquake, which also occurred in 2010 and also had an M_W magnitude of 7.0, was responsible for millions of injuries and hundreds of thousands of deaths.

The vast majority of earthquakes affecting the Earth have tiny magnitudes, 1.0 and less, and we normally don't feel anything when the quake registers less than about 2.5. The largest measured example remains the M_W 9.5 event in 1960, south of Santiago, Chile, not far from the 2010 (magnitude 8.8) and 2015 (magnitude 8.3) earthquakes that did so much damage to the vineyards and wineries in the Elqui, Limarí, and Choapa valleys. All these Chilean examples were mega-thrust earthquakes.

The North Anatolian strike-slip fault runs for more than 1200 kilometers across Turkey (the country with the fourth largest area under grapevines). It crosses the Diren vineyards north of Tokat in the east, continues westward through eastern Marmara, including the vineyards south of Sakarya, before helping to demarcate the great inlet of the Marmara Sea, with Istanbul and Thrace to the north. Very worryingly, in recent centuries earthquake activity along the fault seems to have been moving progressively westward. The most recent major event, the 1969 Gulf of Ismit earthquake and associated tsunami (yes, tsunamis are not restricted to open oceans) killed tens of thousands of people. Some seismologists believe that the next episode is imminent and will be centered yet further west—right in the Sea of Marmara. Such an event would threaten both nearby Istanbul, the world's fifth most populous city, and the Thrace region immediately to the north, the area of Turkey with the greatest concentration of wineries.

The influence of strike-slip faults on topography is shown in the South Island of New Zealand, as much of the straight coastline on the northwestern side is defined by erosion along the Alpine Fault. Here, the moisture-laden westerly winds have to soar abruptly from the open ocean on the west side of the fault up to the high Southern Alps on the east side, triggering one of the highest rainfalls on Earth: close to 6000 millimeters a year. Where the Alpine Fault strikes inland, toward the north end of South Island, it continues its topographic significance, delineating the eastern boundary of the lowlands of the Nelson wine region and diverging into major splays.

It was displacements within this fault system that resulted in the 7.8 M_W earthquake in 2016, which ruptured a number of vineyards. Some of the fault splays essentially define the wine-producing areas of Marlborough. For example, a southerly fork has given rise to the Awatere Valley, and a northerly splay runs north of Blenheim along the north side of the Wairau Valley. These two important vineyard areas are separated by intervening tougher rocks caught between the faults, rocks that make the distinctive upland zone called the Wither Hills. These are all names familiar to wine lovers. And as the Wairau Fault leaves the coast of South Island, its weakened rocks give rise to a broad inlet. In 1770, Captain James Cook gave it a name in English: Cloudy Bay.

That was exactly one year after another explorer, Captain Gaspar de Portolà, had become the first European to set eyes on the "magnificent estuary" we now know as San Francisco Bay. That same night he camped near a small lake not far to the south. A few years later, a Franciscan missionary named Padre Palóu rested by that same lake and, it being the feast day of St. Andrew, called it *Laguna San Andrés*. Fast forward to 1895, when a geology professor from the university at Berkeley discovered a major fault in the region and named it the San Andreas. He explained that the name came from that little lake (now a reservoir, Figure 7.10a; see Plate 16) because it lay right on the fault. However, some of his peers, knowing his egotistical nature, thought he had another reason. His name was Lawson, *Andrew* Lawson.

Figure 7.10 The San Andreas Fault. (a) San Andreas Lake, south of San Francisco, after which the fault is named (see text). The valley with its elongate lake has been eroded along the weakened rocks of the fault zone. (b) Fermentation vessels at the Wente winery, near Livermore, damaged during the 1980 movements along the fault (see text). (c) Satellite view (from Google Earth) of the San Francisco Bay area indicating some major faults in the San Andreas system.

The San Andreas Fault affects the California wine industry in a number of ways. This complex zone of fracturing extends for nearly 1300 kilometers, northward from its southern end in the Salton Trough on the California–Mexico border, on land, but at an altitude of about 70 meters below sea level and one of the very few places where the fault itself is visible (Figure 7.11). It continues northward past Los Angeles, for example separating the Sierra Pelona and Antelope Valley AVAs. If you drive eastwards along Route 14, you cross the fault where the high, hilly scrubland of the Sierra Pelona abruptly gives way to the flatter, high desert of the Antelope Valley (both areas having several wineries). The fault itself lies beneath the alluvium at the foot of the descent; the contorted strata it has produced are spectacularly visible in roadcuts both sides of the highway as it meets the southern outskirts of Palmdale.

Figure 7.11 Part of the actual San Andreas Fault in bedrock, in the Mecca Hills, Salton Trough, southern California. Note the intensely shattered nature of the rock, causing weakness that is readily exploited by erosion vertical face, about three meters across.

The San Andreas fault impinges on much of the Central Coast AVA by affecting rock types and topography. For example, some growers in Paso Robles west of the fault cherish the presence of limestone patches, a rock almost absent from east of the fault. Most of the mountain ranges in California run roughly north–south, but some to the west of the fault have become rotated by the faulting to more of an east–west orientation, allowing cooling breezes to blow in from the Pacific. Burgeoning areas such as the Edna, Santa Maria, and Santa Ynez valleys owe their success to this situation: even this far south, Burgundian cultivars flourish.

Northward, after running along the eastern edge of the Santa Cruz Mountains AVA, the fault and its numerous splays define the shape of San Francisco Bay

(Figure 7.10c; see Plate 16), then head up to Point Arena, on the Pacific Coast west of Anderson Valley and Ukiah, from where the structure heads out to sea. The fault with all its subsidiaries reactivates from time to time and hence generates earthquakes. The headline-making Napa earthquake of August 2014 (M_W magnitude 6.0) was due to renewed slip on the West Napa Fault, which is a splay of the Carneros Fault, a subsidiary of the Hayward-Rogers Creek Fault that underlies the Sonoma Valley and that itself is a splay of the San Andreas Fault proper. These are complex networks!

Most of the fault's course is characterized by valleys, but slivers of tougher bedrock can give localized hills: Hirsch Winery in northern Sonoma County and Savannah-Channelle and Cinnabar vineyards in Santa Clara County are on low hills within the fault zone. DeRose Vineyards, south of Hollister, is sited right on the San Andreas Fault. A crack in the winery's main building continues to open about a centimeter each year and runs on through the warehouse between the fermentation tanks and the aging barrels. At Wente Brothers winery south of Livermore, the Greenville splay of the San Andreas Fault moved in 1980, overturning and rupturing many of the stainless steel tanks there, leaking around 90,000 liters of wine. Some of the tanks were righted and refurbished, but they still show striking crumples (Figure 7.10b; see Plate 16).

Unsung but Ubiquitous: Joints in Rock

In contrast with the much heralded structures just discussed, joints in rock are little heard of. Yet we have all seen them, and in their low-key way, they exercize enormous influence. Rock **joints** are not places with loud, driving music, but systematic arrays of small fractures criss-crossing bedrock. However, unlike faults, there is virtually no displacement across them. Every rock face you may have looked at, a road cutting, crag, or cliff face, will have these joints, but, rather like the lichens that grow on rock, we tend to subconsciously filter them out and not notice them. The reason for the ubiquity of joints is that rocks can't resist much pulling apart without breaking—and most rocks now at the Earth's surface have been buried at the very least by a small amount. When they are exhumed, and the burial pressure is taken away, they relax—rather like a portly person taking off a corset—and expand a tiny bit in all directions. Being very brittle at the Earth's surface conditions, they fracture to make joints.

Typically, in a pile of more or less horizontal sedimentary strata, there will be joints roughly parallel to the bedding and a couple of vertical arrays (Figure 7.12a). In the two-dimensional section of a cliff face, both vertical sets will appear as one, but if we look down on a bedding surface, we will see the two different sets. The spacing of the joints roughly varies with the thickness of the beds: thicker beds give more widely spaced joints. No rock face is completely smooth, and in detail the

Figure 7.12 Examples of joints in rock. (a) Sandstones in Caithness, Scotland. Two sets of vertical joints approximately at right angles to each other criss cross the rock and, together with a tendency for the rock to split parallel to the horizontal bedding, induce the rock to break into square slabs. (b) Columnar joints in basalt, due to contraction during cooling. Hegyestű, near Monoszló, Badacsony wine region, Hungary.

irregularities in its form probably coincide with the joints. Being fractures, they are relative weaknesses, and, as with faults, erosion picks on them.

If you stand in the vineyards above the village of Solutré-Pouilly, in the Mâconnais, France, and look up at the inclined beds in the famous Roche de Solutré, its form is partly dictated by the strata but partly by the joint sets. In the lower, slightly red-weathering parts, the beds are relatively thin, the joints close together, and the rock face has a decidedly irregular, almost rubbly look. This contrasts with the thick limestone layer at the top of the cliff, where the widely spaced joints give an angular but smooth look to the face (Figure 7.13; see Plate 17).

Underground waters, often with dissolved material in them because of the slightly increased temperatures there, tend to percolate along joints. If the water encounters a change in conditions, such as a temperature drop, the solubility may fall and any dissolved minerals will precipitate out. Irregular but vaguely sheet-like bodies of minerals form in the joints, known as **veins**. They can form in any kind of fracture and are common in faults, but they normally bear no relation to other features in the surrounding rock, such as stratification or foliation (Figure 7.14).

Common parlance may talk about veins of, say, limestone or schist, below a vineyard, but geologically this is incorrect: veins are localized zones of minerals that cut across whatever the host rock may happen to be. Most veins aren't made of rock. They can comprise all kinds of precipitated minerals, including some that are of tremendous interest in commercial mining, but veins are typically filled by either calcite or quartz. So those white gashes you so commonly see in boulders or outcrops of rock are more than likely mineral veins. When weathering attacks these rocks, pieces of the vein material may fall out, and many will be transported by a river down to the sea. Those white pebbles conspicuous on an otherwise dark-colored river bank or shingle beach, and in many vineyard soils, probably originated somewhere in a bedrock vein.

A special kind of joint arises when igneous sheets like sills cool and contract. The freshly solidified rock breaks into joints that can have a distinctive polygonal shape,

Figure 7.13 Example of joints in rock: the Roche de Solutré, overlooking Solutré-Pouilly in the Mâconnais, southern Burgundy, France. In this view, the bedding (stratification) is slightly inclined downward towards the right. The differing toughnesses give a general staircase-like form to the cliff (see Chapter 8), but the shape of the rock face is largely governed by the sets of almost vertical joints.

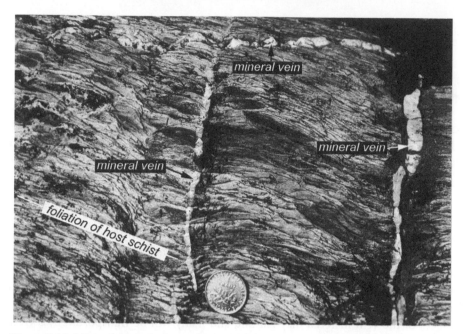

Figure 7.14 Mineral veins. In this instance, the foliation of the host schist (these terms are explained in Chapter 6) is slightly inclined downward toward the right of the photograph but is cross-cut by two sets of mineral veins. One, near the top of the photograph, is horizontal, and the other is vertical; it's their lack of physical relationship with the surrounding rock that makes them veins. Both sets were originally fissures developed along joints in the host rock that were later infilled by precipitation from percolating waters—in this example by quartz.

reminiscent of angular organ pipes. Such **columnar joints** are found only in igneous rocks, but they can be spectacular. Well-known examples include Devil's Tower in eastern Wyoming (of "*Close Encounters of the Third Kind*" fame) and Fingal's Cave on the Isle of Staffa in the Scottish Hebrides (of Mendelssohn overture fame). Regarding vineyards, they range across the basalt cliffs overlooking the vineyards on Somló and Badacsony Hills, Hungary (Figure 7.12b), and form the Palisades above Calistoga in the northern Napa Valley, California.

Joints can become crucial where vine roots attempt to penetrate bedrock, seeking additional water. As the main roots grow down through the soil, they are preceded by myriad tiny hairs capable of sensing increasing water presence and signaling this presence to the vine. The main roots are then induced to extend toward the water. If necessary, they can exert a pressure (almost five times greater than that in a car tire); thus, provided this exceeds the strength of the soil, the roots will probe further. If they meet intact bedrock, the root forces will not likely be sufficient to penetrate that, but where the root hairs meet a joint (and their diameters will be orders of magnitude less than the width of the fracture), they may well sense the presence of water stored down within it and trigger the main axes of the root to exploit this plane of weakness.

In parts of the Colchagua Valley, Chile, the alluvial pebbles are sufficiently weathered and crumbly for vine roots to be able to grow through them. However, they store little water, so where there are clay-rich layers underneath, the vine roots grow right through the gravels in order to access the water stored in the underlying clays. Consequently, they can grow downward to a depth of several meters or more. By contrast, not far away in the Leyda Valley to the north, some vine roots grow to less than a meter before hitting granite bedrock, which is sparsely jointed and impenetrable to the roots. Machinery can be used to rip and loosen the soil as a means of easing root penetration, but it's little match for intact bedrock granite. (As I explain in Chapter 9, this is all about pursuing water. Phrases involving deep roots seeking water and minerals/nutrients are commonplace, but little nutrition will be available in these sterile, intact materials.)

Chapter 6 explained how the preferred vineyard sites of the Douro region of Portugal are accounted for by the steeply oriented foliation in the underlying metamorphic rocks. as opposed to the granite areas. Like all rocks, the granite has joints, and some of these are also steeply oriented. However, in such a massive rock, completely lacking in bedding and the like, the joints are spaced on the scale of meters, much too far apart to be of use to the vine roots and to accommodate much stored water. In such a very dry situation, the bedrock joints are not only influencing grape quality, but also they can be the difference between life and death for a vine. So clearly joints can be crucially important in vineyards. Moreover, these little heralded structures are routinely exploited by quarrymen, civil engineers, mining geologists, water geologists, and others. They are everywhere beneath our feet, quietly, in one way and another, exerting their unsung influence.

8

The Lay of the Land

Every farmer knows that certain crops do better in particular fields, and every gardener knows that some plants grow better in certain spots in the garden. Grapevines are no different. The idea forms the basis of the concept of terroir, and in this and the following two chapters we will meet a number of factors, besides the minerals and rocks we have been talking about, that contribute to it. First, we consider the shape of the land surface.

The weathering of rocks produces loose debris—sediment—which sooner or later will move, and this gives rise to **erosion**. The two processes usually work hand in hand though, strictly speaking, weathering happens in place, whereas erosion results from *movement* of the debris. We will look more closely at weathering in the next chapter, in the context of generating soil. Here we are concerned with erosion. It may involve sand particles being hurled at outcrops by high winds, or rivers loaded with particles grinding at the land to form a river channel. In some places, rock-charged ice may be gnawing away at the bedrock. Ultimately, the shape of the land surface is the result of how such processes interact with the solid bedrock. In other words, the interplay between erosion and bedrock determines the physical lay of vineyards.

Sculpting the Land: Some Features Due to Erosion

Plateaus are level upland areas. They can be formed in any kind of material: it's the flat, table-like form that defines them. For instance, a vast area of the Deccan Plateau of central India, focus of a burgeoning wine industry, is made up of horizontal flows of basalt lava. The Colorado Plateau in the United States is formed largely of horizontal sedimentary strata. It has been deeply incised by rivers, in places leaving isolated blocks such as **mesas** and **buttes** (Figure 8.1; see Plate 18). Mesas have a larger summit area than buttes, compared to their heights. These bodies of rock have not been individually uplifted, as is sometimes claimed. They are remnant blocks, erosion having taken away the strata that were once around them.

Figure 8.1 Example of a butte: Mount Garfield, Mesa County, Colorado. The sandstone strata of Book Cliffs overlook the several wineries sited on East Orchard Mesa, between Grand Junction and Palisade. Modified with permission from an original photo by Denise Chambers/Miles, Colorado Tourism Office 2015.

Although the Colorado vineyards that are these days flourishing to the west of the Rockies are mainly sited on the plains, they are surrounded by mesas and buttes that lend their names to some of the wineries and their products. Just over the state line to the west, even Utah now boasts wineries among its famous red rock buttes and mesas; in fact, these two terms are associated with wineries all across the American Southwest. Elsewhere, Cape Town's Table Mountain, with the historic Constantia vineyards snuggled below it, is technically a mesa, though rarely referred to as such. The vines on the Butte Montmartre are a well-known Paris landmark; a number of hills in the Loire Valley are termed buttes, such as the location of the town of Sancerre and the Butte de la Roche in the western Muscadet region; Chouilly in the Côtes des Blancs district of Champagne has the Butte du Saran.

Where a mesa-like block is formed from gently *inclined* strata of sedimentary rocks, we call it a **cuesta**. Look for examples of all these features the next time you're watching a classic Western film. Not surprisingly, the term *cuesta* is used in Spain, finding its way into a number of wine names there (Finca la Cuesta and Cuesta sherry, for instance) and in France. The Coteaux de Saumur and part of adjacent Saumur-Champigny are described in French as lying on a cuesta. Cuestas—even if they're not referred to as such—are tremendously important landforms for vineyards, as they give both a long, even slope lying parallel to the inclined bedrock—the

dip slope—and a relatively narrow, steep side that has been cut down across the strata, called an **escarpment** or **scarp**. The Weald-Artois anticline mentioned and mapped in Chapter 7 gives rise in England to the south-facing scarp slopes of the North Downs, now home to nearly a dozen vineyards between Dorking and Aldershot alone. To the south, the nearly two dozen vineyards between Hailsham and Petersfield are on the equivalent rocks of the South Downs, again on south-facing slopes but here on the southern limb of the region's anticline, effectively the dip slopes of a cuesta.

Some vineyards are located in sheltered sites next to escarpments. For example, fundamental in northeastern North America is the Niagara escarpment, which arcs northeastward from the Illinois–Wisconsin state line to help define the northern shores of Lakes Michigan and Huron, eastward through the Niagara Peninsula, and well into New York State. Its tough dolomite accounts for Niagara Falls and frames a number of wine-producing areas such as the Niagara Escarpment and Wisconsin Ledge AVAs. In Canada's Niagara Peninsula VQA, the escarpment helps contain the climatic interaction between the vineyards on its slopes and adjacent Lake Ontario.

The term *escarpment* is also used more broadly for any relatively steep, rather isolated slope, irrespective of the bedrock. For example, the escarpment that defines the southwestern boundary of the Walla Walla Valley AVA in Washington is developed in basalt. In French, an escarpment is sometimes referred to as a **côte**, which also can be formed in any kind of bedrock. The Côte Rôtie, for example, is granite, and around Montreal, Canada, some of the vineyards involve côtes developed in unusual and commercially important igneous rocks. Analogous escarpments due to geological faults—most famously France's Côte d'Or—are discussed in Chapter 7.

Where the bedrock strata are steeply inclined, such that the dip slope of the escarpment is about as steep as the scarp slope, the landform is known as a **hogback**. Again, although normally developed in sedimentary strata, hogbacks can be eroded in other rocks, such as tilted lava flows. The name appears in the Columbia Gorge AVA of Washington and Oregon and in winery names there. The example near Guildford in the chalk of England's North Downs, now with vines growing on it, is known by its time-honored name of *Hog's Back*. Jane Austen called it that.

As erosion proceeds, any particularly resistant bedrock masses become increasingly isolated. Totally lone lumps, particularly of nonsedimentary rocks, are sometimes referred to as **monadnocks** (a Native American word), especially in North America, or **inselbergs,** especially in Africa. Granite monadnocks are striking in the Texas Hill Country wine region, especially around the town of Fredericksburg, and there are numerous inselbergs in the Western Cape region of South Africa. The upland mass of Fruška Gora, with numerous vineyards dotting its slope, sits isolated in the plains of northern Serbia and is sometimes referred to as an inselberg.

Erosion can also gouge out the land surface to leave low-lying areas, which may be filled more or less temporarily with water: **lake basins**. These range in size from the local, such as the one at Caldaro al Lago (Kaltern am See) in Italy's Alto Adige

DOC, to the vast Qinghai Lake Basin of west-central China. I discuss the modifying effect of bodies of water on vineyards in Chapter 9.

The Influence of Bedrock

The Kind of Rock

Bedrock influences landform through its resistance to erosion relative to adjacent rocks. Given the variability of rocks, it's risky to generalize, but here goes (Figure 8.2). Most igneous rocks are tough because of the interlocking arrangement of the minerals produced by cooling of the parent magma. Metamorphic rocks are much more variable: slate, schist, and marble in general tend to be weak. Gneiss tends to be much more resistant to weathering, and generally just about the most durable rock of all is quartzite, with its grains of tough quartz firmly bonded together. The resistance of sedimentary rocks depends partly on what they are actually made of—varying from weak, clay-rich rocks like shales and marls to the tough quartz of sandstones—and partly on how well the constituent grains are bonded together. Many sandstones are relatively tough, but where the grains are poorly cemented together, they can be noticeably weak and crumbly.

In southern Indiana in the United States, well-cemented sandstones and relatively durable limestones have resisted erosion to produce higher land with a climate slightly different from surrounding areas. Also, because this upstanding area escaped being covered by ice during the recent glaciation, its soils are developed in loess rather than till. All this is the basis of the Indiana Uplands AVA, that unusual thing—an *appellation* fundamentally delimited by geology.

One kind of rock gives rise to particularly distinctive features: limestone. As rocks go, limestone dissolves easily, which gives rise to a series of characteristic landforms (Figure 8.3). Such scenery is referred to as **karst** and the features as **karstic**. The name comes from the area of southwest Slovenia and northeast Italy known as *kras* and *carso* respectively. Early publications on its distinctive scenery were in German and called karst, and this has been adopted in English. Most of Slovenia's Teran wine comes from the classic karst region. The French word for a plateau made of limestone is a *causse*, well known in southern France. Cahors, for example, distinguishes between its riverside vineyards and those located up on the calcareous plateau called simply "Les Causses." And, of course, limestone dissolved below ground can give rise to systems of **caves**, with their obvious importance for wine cellars.

The Shape of the Rock

The shape of a rock mass can be important where the rock is significantly tougher than the materials round about. South Africa's striking Paarl rock (an inselberg) reflects the form of a granite pluton; cone volcanoes (Chapter 4) develop their

	ROCK NAME	RESISTANCE	LANDFORMS
IGNEOUS *FINE GRAINED* DARK (mafic)	**BASALT**	Usually resistant except when bearing olivine. Exfoliates readily	Columnar jointing, narrow ridges from dikes, escarpments from sills
PALE (siliceous or felsic)	**RHYOLITE**	Usually resistant, but some examples weather readily	Bluffs and cliffs
MEDIUM GRAINED DARK (mafic)	**DIABASE/ DOLERITE**	Usually very resistant except where much jointed and where containing olivine	Ridges from dikes; escarpments from sills
PALE (siliceous or felsic)	**ANDESITE**	Usually resistant	Not widespread enough to form typical landfroms
COARSE GRAINED DARK (mafic)	**GABBRO**	Usually very resistant except where much jointed and where containing olivine	Ridges from dikes; escarpments from sills
PALE (siliceous or felsic)	**GRANITE**	Usually resistant but some examples weather readily	Monadnocks, inselbergs, uplands; exfoliation domes
SEDIMENTARY *FINE GRAINED* (argillaceous) SEDIMENT	**CLAY, MUD**	Weak but usually cohesive enough (unless wet) to form vertical walls	Bluffs and badlands
ROCK	**SHALE/ MUDSTONE**	Usually weak	Gentle slopes, valleys and lowlands
FINE GRAINED (calcareous) CLAYEY	**MARL**	Very weak	Low valleys, very gentle slopes
ROCK	**LIMESTONE**	Weak in humid regions; resistant where arid	Karst. Cliffs, high escarpments
COARSE GRAINED SEDIMENT	**SAND**	Usually weak, especially where dry or very wet	Lowlands but can cap uplands
ROCK	**SANDSTONE**	Usually resistant, depending on degree of cementation	Cliffs and plateaus
VERY COARSE GRAINED SEDIMENT	**GRAVEL**	Moderately resistant, depending on nature of pebbles	River beds, can cap uplands
ROCK	**CONGLOMERATE**	Very resistant	Ridges and mountains
METAMORPHIC	**SLATE**	Weak, but often more resistant than limestone	Lowlands
FOLIATED	**SCHIST**	Variable but often moderately resistant	Uplands and ridges
	GNEISS	Usually very resistant	Uplands
NONFOLIATED	**MARBLE**	Weak	Lowlands, unless with more resistant metamorphic rocks
	QUARTZITE	Very resistant, perhaps the most durable rock	Ridges, monadnocks, inselbergs

Figure 8.2 Generalities on the relative resistances to erosion of different rock types and the resulting landforms.

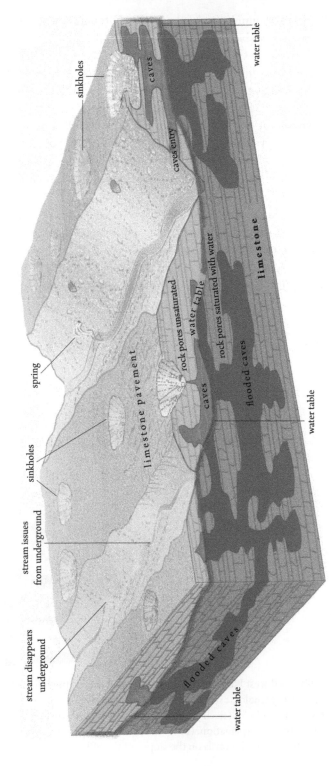

sinkholes

caves

water table

caves entry

rock pores unsaturated

water table

rock pores saturated with water

limestone

caves

flooded caves

water table

spring

limestone pavement

sinkholes

stream issues
from underground

stream disappears
underground

flooded caves

water table

Figure 8.3 Block diagram of features associated with limestone bedrock.

shape through the local accumulation of the relatively tough material they eject, and, of course, many of them have vineyards on their slopes: the Garrotxa area in Catalunya, Spain; Santorini in Greece; Etna, Vulture, and Vesuvius in Italy.

In Hungary, when the sediments that eventually were to form the Carpathian Mountains were accumulating, their weight caused the area to the south, now the Hungarian plain, to become stretched, and in places volcanoes broke through. They formed a series of craters, at first filled with tephra. Their rims of ash were able to hold in the flows of basalt lava that followed, and although these borders were subsequently eroded away, this left an array of flat-topped and distinctly isolated basalt hills. Today they rise from the Hungarian plain, in places like Somló, Hungary's smallest and oldest wine district (Figure 8.4; see Plate 19).

In a stack of sedimentary strata, the individual layers will almost certainly show differing resistances to erosion, thus giving a staircase form to the face of many an escarpment. A much photographed example is the Roche de Solutré in the Pouilly-Fuissé district of the Mâconnais, France (Figure 7.13). Here, jutting out above the clays and marls below the cliff is a series of limestone strata, some tougher than

Figure 8.4 Somló Hill, northwest Hungary. (a) The isolated nature of the hill, its lower half clothed in vineyards. (b) Geologic map. Most of the area consists of sands and gravels (e.g., Qp4 and Qph3), but Somló Hill itself is largely determined by extrusive basalt (MPl3). See the text for further explanation. (c) Loose blocks of the basalt being used to build heat-retaining walls next to vineyards on the slopes of the hill.

others and so giving a staircase form, capped by a thick layer of particularly dense and tough limestone.

Weathering and erosion open up rock joints into fissures, and their walls become progressively more separated. This is the reason that mesas and buttes tend to be so steep-sided; their edges are defined by joint systems. Joints contribute to the well-known rugged forms of a **badland** landscape, and erosion may progress until only isolated pinnacles remain, as in the skyline of a number of wineries in and around the Santa Lucia Highlands AVA, California. In limestone, joints criss cross to give a pattern known in English as a **limestone pavement** (Figures 8.3 and 8.5 [see Plate 20]). As long ago as 300 B.C., Theophrastus of Lesbos noted that particular shrubs grew on limestones, and in southern France, the *garrigue*, with its herby fragrance sometimes mentioned in wine tasting, is helped by the shrubs gaining protection from sun and wind by growing in such joint fissures. Similar scrubland developed on nonlimestone rock is known as *maquis*.

Folded strata can have a major effect on the landscape, as we have just seen for southern England. The accumulation of river and lake sediments in the Bekaa (Beqaa) Valley in Lebanon, where most of the country's wineries congregate, is founded on a major, open syncline (Figure 8.6) with the flanking mountain ranges formed on anticlines, all accentuated by faulting. Figure 8.7 (see Plate 21) gives examples of strata deformed into tight complex folds that have a major effect on landforms and hence on vineyard distribution.

Chapter 7 explained how geological faults affect the landscape. As further examples: in the Bas-Beaujolais, faults have juxtaposed sedimentary rocks with differing durabilities to give jostled, fault-bounded blocks, each of slightly differing altitude.

Figure 8.5 Limestone pavement, West Yorkshire, England.

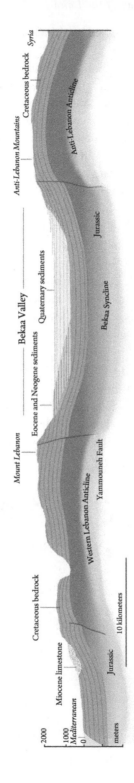

Figure 8.6 Sketch cross section of the Bekaa Valley, Lebanon, and adjacent mountains to show their relation to folds in the underlying bedrock.

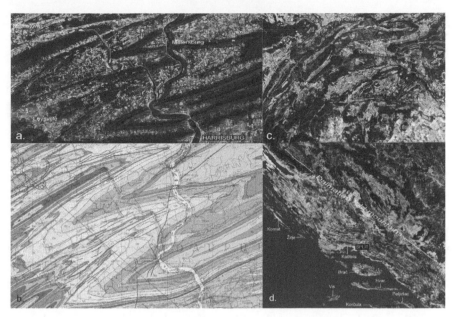

Figure 8.7 The effect of folded strata on landform. (a) Arcuate ridges (forested) and valleys in the Appalachians north of Harrisburg, Pennsylvania. (b) Geologic map showing the patterns typical of highly folded strata, here of Ordovician to Triassic age, in the Central Pennsylvania winery region approximately coinciding with the area shown in (a). The grid squares are 10 kilometers across. From the 1980 Geologic Map of Pennsylvania. (c) Arcuate forested hills following highly folded bedrock strata southeast of St-Chinian, France. (d) Highly folded strata around Split, Croatia, making arcuate ridges on land and a series of linear offshore islands. All the islands named have a winemaking tradition; Kaštela is the location of the original Zinfandel vine. Figures 8.7a, c, and d are based on Google Earth images.

In the United States, the Balcones Fault system has brought the tougher rocks that underlie the Texas Hill Country AVA next to the softer rocks that underlie the Austin and San Antonio plains, giving an escarpment used for viticulture (and giving its name to Texas's first modern whisky—thoroughly Texan in style but, unlike other American examples, spelled whisky). In Australia, the stepped appearance of the southern slopes of the Mount Lofty Ranges, which include the Adelaide Hills wine region, results from a series of faulted blocks. In such ways, it's the nature of the bedrock geology and the form in which it is arranged that, ultimately, goes a long way in determining where vineyards are sited.

The Sediment Settles: From Ice, Wind, and Water

When the currents causing erosion slow down, the debris they are carrying may begin to settle. If this comes about through melting ice, the resulting till commonly forms a ridge-like deposit known as a **moraine**. Some of the vineyards in Vaud, Switzerland, the Niagara Peninsula VQA in Ontario, Canada, and in Central Otago, New Zealand, are sited on moraines. Consequently, the soils show a large variation of fragment sizes mixed in with clayey material. In New York State, some glacially carved valleys became temporarily dammed by moraines as the melting ice retreated, giving rise to the Finger Lakes, and moraines define the form of Long Island, both now AVAs.

In major contrast to this disorganized dumping of sediment from melting ice, material being deposited from wind or water is subject to tight physical controls. They're summed up in Figure 8.8. It's an important graph in a number of fields of study, yet it first appeared in an unpublished student thesis, that of Filip Hjulström at Uppsala University, Sweden. It may look a bit technical, but essentially it shows that as air or water moves faster, it's capable of moving larger particles, and vice versa. Torrential flow might even be capable of moving boulders, as in the chasm in Coleridge's *Kubla Khan*, where in the sacred river "huge fragments vaulted like rebounding hail." Such coarse particles, and down to sand in grain size, are dragged by the stream on or near its bed in what is termed the **bed load** of the river. The size of the particles being carried by a river isn't fixed, though, because of abrasion (Chapter 5). Incidentally, such coarse fragments are called *sassi* in Italian and give their name to wines such as Campo Ai Sassi and Sassicaia.

Silts and clays, as Figure 8.8 shows, are moved by much slower rivers, lifted and carried in **suspension** by the turbulent action of flowing water, sometimes for long distances. (We met wind-carried silt—loess—in Chapter 5.) When a river slows down, the particles may cease to be dragged, or they may settle out of suspension;

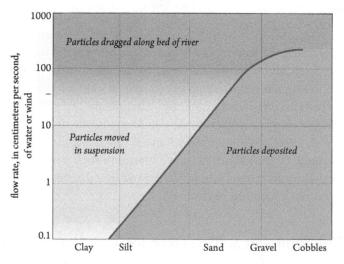

Figure 8.8 The "Hjulström curve," showing how particles of different sizes are moved by wind or water in suspension or along the floor, or are deposited, according to their size.

that is, they will be deposited. Thus, because river flow is constantly changing, at any place the particle size being deposited can vary abruptly through time. Also, of course, rivers change their course. This is the main point for us: vineyard soils on the sites of old river courses will show rapid changes in arrangements of particle size and hence drainage. Consequently, with every few paces you take in such distinguished places as Rutherford (Napa, California), Pauillac (Médoc, France), or Marlborough (New Zealand)—all on river deposits—the drainage properties of the soils change markedly.

All these river sediments are conveniently described as **alluvial**, and the material is **alluvium**. I occasionally see the word "diluvium" in wine writings used as though it were a synonym for alluvium. In geology, even when diluvium was in its early nineteenth-century heyday, the word had nothing to do with rivers, and it's now wholly archaic—well, it's antediluvian. One final point about settling sediment: we mustn't forget that some weathered material is removed by having been dissolved. It can be carried great distances in solution, but changes in physical and chemical conditions or biological activity may eventually cause **precipitation**. Precipitated sediments can occur locally, and they can veneer wide areas of the land surface, such as in dried-up lakebeds (Chapter 5). Of course, much of the material dissolved in river water will be carried all the way out to sea—after all, that's why the oceans are salty.

River Valleys

There are notable wine-producing areas in valleys carved by glaciers—the Valle d'Aosta and Alto Adige in northern Italy, for example. Nevertheless, if you think of the world's great vineyards, you may well picture river valleys. They may be in the New World: Napa, Casablanca, or Barossa perhaps, or in the Old: Loire, Rhône, Douro, or Rhine, for example. We therefore now focus on river valleys, beginning with some generalities about rivers flowing in their upper reaches.

Rivers in the Hills

As rivers flow from their source in higher land, they lose altitude relatively quickly at first and then fall more slowly as they approach the sea. Generally, the overall shape is concave (Figure 8.9) though this shape is less apparent in exceptionally long rivers such as the Mississippi or the Orange, in South Africa. Normally, the steeper upper parts are characterized by erosion, with the river cutting vertically downward to give a **gorge** or **canyon**. In most cases, the sides that define a gorge will be too steep for vineyards, but they are vulnerable to erosion and downslope slip (see hillslopes, below), which through time reduce their steepness and open up the gorge. The river is still downcutting, and so we get the **V-shaped valley** that is so characteristic of the upper parts of river valleys. Viticulture may be feasible here, but

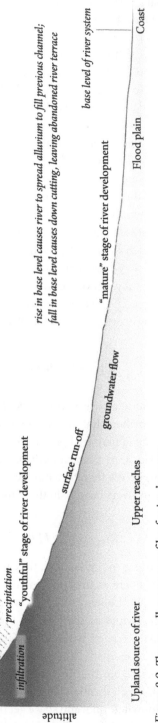

Figure 8.9 The overall concave profile of a river's course.

the steep slopes will be a challenge. Anyone who has visited the Mosel and Douro valleys, say, will have been struck by the spectacular vineyard terraces, patiently constructed to make vine growing possible on these steep slopes.

Gorges occur *within* the dominantly V-shaped course of a river where the rate of downcutting is locally exceeding the slope-lessening processes. This comes about if the land surface is made higher, as with the accumulations of lava in the Columbia Gorge AVA in Washington and Oregon. Or the river may encounter an area of tougher rock that allows some downcutting but that resists downslope movements. A classic European example is the legendary Lorelei rock in the Rhine Gorge— a steep, craggy, vineless stretch between the vineyards of the adjacent somewhat more open valley. The district as a whole was raised in geologically recent times, resulting in the gorge of the Middle Rhine. It's carved mainly in slaty rocks, but at Lorelei the river cuts into a particularly tough sandstone.

In more arid regions, streams may only flow after heavy rainfall: a temporarily dry river valley is referred to in Spanish as an **arroyo**, a term common in the Californian wine world. These dry valleys usually have a rock-strewn flat floor that contrasts with the V-shape found in more humid areas.

Out on the Plains

A marked change in character comes about as the river continues on its seaward course. The river levels out and downcutting is reduced; slope processes come to dominate, and the opposing sides of the valley separate more and more. The valley really opens up. The volumes of river water are greater, as is the amount of suspended sediment. Any deposition that takes place along straight stretches of the river's course may build a natural **levee**. But the river channel will be constantly changing, and its path less straight, giving the lazy, looping, muddy river typical of lowland areas.

The form is classically displayed just north of Sarayköy in western Turkey. Here the river breaks out of the narrow mountain gorges to snake its languid way across a broad, flat-bottomed valley toward the Aegean Sea. Its winding route is striking and was celebrated in classical times: the river is mentioned in Homer's account of the Trojan Wars, as well as in the writings of Aristotle, Strabo, and Xenophon The river's name? Transliterated from the Greek, it's *Maeander*, a word that has, of course, come down to us to mean rambling and is also the technical term for the pronounced loops commonly seen in the lowland paths of rivers. Mark Twain offered that the Mississippi River was being shortened by all its loopings, rather like the curls in a slack piece of string. He therefore figured that before the meanders formed, the river must have been "stuck out over the Gulf of Mexico like a fishing-rod." That's Mr. Twain for you.

Erosion of the banks is greatest at the outer bend of a river curve, sometimes called the **cut bank**, and sedimentation takes place at the inner bend, producing a **point bar** (Figure 8.10). Exactly how this happens isn't quite as simple as it might

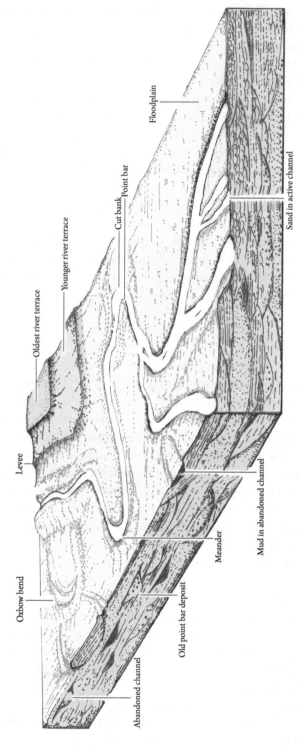

Oldest river terrace

Younger river terrace

Cut bank

Point bar

Floodplain

Levee

Sand in active channel

Oxbow bend

Meander

Mud in abandoned channel

Abandoned channel

Old point bar deposit

Figure 8.10 Sketch block diagram to show some lowland, downstream features of a river and the complex arrangements of sediments that result. Such rapid variations of soils are typical of vineyards in an alluvial setting.

appear: one of the first scientists to tackle the problem was none other than that indefatigable thinker, Albert Einstein. The simultaneous erosion and deposition processes act to progressively increase the sinuosity of the river and hence produce the **meanders.** There are particularly fine examples of these features in France's Atlantique IGP region, such as on the River Lot in the Cahors AOC (Figure 8.11). The district's annual wine fair is held in the little village of Parnac, huddled within one such meander, and the old town of Cahors itself is in another. Just downstream, the village of Luzech sits in an extreme meander, almost a complete river loop, called an **oxbow bend.** Up on the Vézère, there's such an oxbow immediately north of Marzac (a mere 3 kilometers north of where the remains of Europe's earliest human were found, in a little cave known locally as a *cro* that belonged to a Monsieur Magnon). An example of an oxbow on the Napa River, California, gives its name to the eponymous district of central Napa town and its vibrant food market.

A river may lack meanders but instead have a relatively straight course containing within it an ever-changing network of minor channels, shoals, and gravel islands. This is a **braided** river. In an intricate interplay between gravel deposition at times of low flow and erosion in time of flood, channels split and switch, and the river channel moves back and forth. The Canterbury Plains of New Zealand's South Island are famous for their braided rivers—the Waipara, for example, with its scores of flourishing wineries, and the Waimakariri, where one winery is called "Braided River."

All this constant changing, this splaying of channels, migration of meanders— across the valley bottom and beyond—gradually broadens the valley and its spread of sediment to develop a **floodplain.** The sediments here tend to be well drained and hence good for vine growth, while the flatness is ideal for mechanized agriculture. It's because some of the world's most extensive vineyards are located on floodplains that the channel complexities are significant. The soils may at first look uniform, but it follows from the reasons outlined earlier, when we looked at the Hjulström diagram, that they will vary abruptly in physical properties and especially drainage behaviors.

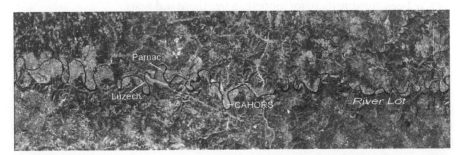

Figure 8.11 Meanders on the River Lot, southwest France. The western two-thirds of the area is in the Cahors AOC region.

In his book *Wine Science*, Jamie Goode reports on PV (precision viticulture) studies on the floodplain of the Wairau River in the Marlborough region of New Zealand. Changes from 2-meter-thick silts to highly stony soils were found within 20 meters of each other because of past changes in the river's course, and links were demonstrated between these and vine performance. Such rapid changes in sediment nature also affect the roughness of the ground surface as well as water availability. Other detailed PV studies have shown that the resulting variations in growth vigor can occur down to a vine-to-vine scale.

Most river systems carry some sediment to their extremity, the **estuary** where they finally empty into the sea. Vineyards are found on the banks of present-day estuaries—the Gironde at Bordeaux is an obvious example, and one winery on New Zealand's North Island is named after such a feature. In England, the Exe and Dart estuaries have vineyards, as does the Bear River in Nova Scotia. Sea levels have varied in the recent geological past, so that buried estuarine deposits can now be found some way upstream, such as at the Wairau River, Marlborough, New Zealand, and the Swan Bay peninsula, Tasmania, which juts into the upper reaches of the Tamar estuary.

If substantial deposits are able to accumulate at the river mouth, they commonly take on the triangular shape of a **delta**. Vineyards on deltas include those where the Waihopai River meets the open Pacific in South Island, New Zealand, and in the so-called delta region where the Sacramento River empties into an inland arm of the Pacific. The vines still growing out on the Camargue Delta of the Rhône are unusual in France, as the salty water and sands at their roots enabled them to survive the phylloxera crisis. We usually think of deltas as forming where rivers meet the sea but triangular-shaped accumulations of sediment can form in other situations. The Hanging Delta in the Finger Lakes AVA of New York formed where a river emptied into a glacial lake that was present at the time, and the Delta vineyard area of Colorado exists where two rivers meet. The term *delta* comes from classical times, when Herodotus likened the outpouring of the Nile into the Mediterranean to the Greek letter for D. Grapevines grow on the Nile Delta today, though they're not native to the area, having been brought in. They're hardly newcomers, though; they've been growing there for more than five thousand years.

Three Favored Terrains

Alluvial Fans

As a river leaves the hills and begins its downstream journey, its channels are no longer confined. Suddenly they are able to diverge out into the main valley, prompting the sudden disgorging of their sediment loads to create an **alluvial fan**. Mountain streams tend to be fast flowing, but they suddenly lose speed upon entering the main valley. Hence, the first material laid down is usually very coarse. Gravels, perhaps including boulders, are dumped at the narrow, uppermost point of the fan, called its **apex**. But they soon block the channel and the water is forced to switch to new courses, and a conical form takes shape, with the bulk of the deposits spreading out over the **apron** of the fan. The lowermost spread of the fan is called its **toe**.

These are important features in the wine world. Figure 8.12 shows a part of southwestern France that is often called the Plateau de Lannemezan; actually, it is the coalescence of a number of alluvial fans spreading out from the north flanks of the Pyrenees. Much of the material that now floors the Côtes de Gascogne AOC consists of sands and gravels that started to accumulate in these fans a few million years ago. Sediment is still being added today by the tributaries of the Garonne and Adour rivers. The Lavaux vineyards of Switzerland are located mainly on a series of alluvial fans, forming where rivers emerge from the Alps bordering Lake Geneva.

Some of the great names of the Napa Valley are associated with alluvial fans. As it happens, the features on the east side of the valley tend to be smaller, but even so they are the locations of the celebrated Screaming Eagle, Shafer, and Clos du Val vineyards. On the west side, alluvial fans are where renowned Napa characters such

Figure 8.12 Alluvial fan (the Lannemezan Plateau) on a geologic map of Gascony, southwest France. The dark-colored formations at the bottom (south) of the map are the older bedrock of the Pyrenees. Rivers emerging from the Pyrenees around the town of Lannemezan have shed gravel and sand (pale colors on this map) as they fan out northward toward Bordeaux.

as Gustave Niebaum, Georges de Latour, and Robert Mondavi made wine. They are the sites of preeminent vineyards such as Bella Oaks, Martha's Vineyard, and To Kalon. In wine terms, alluvial fans are important.

In Napa and elsewhere in California, a spread of alluvial sediment between uplands and valley floor has come to be known as a **bench**, a local, informal word not to be confused with lakeside benches or hillslope benches (Chapter 8). Examples include the Kelsey Bench in Lake County and the Silverado Bench of Napa. But the famous one is the Rutherford Bench, manifest as a gentle incline west of Route 29 between the towns of Oakville and Rutherford. It's a rather cryptic feature to look at, so for visitors wanting to experience at first hand the celebrated Rutherford Bench, the Franciscan Estates winery has helpfully installed a suitably carved wooden bench! The soil on the Rutherford Bench is also well known and is dubbed "Rutherford Dust." Stories abound regarding whether it was the fabled winemaker Andre Tchelistcheff or the esteemed academic Maynard Amerine who coined the epithet. And what the term meant. Maybe it was something to do with a gold-dust allegory rather than anything literal because in these alluvial fans, very fine-grained deposits—dust—will be sparse; the Rutherford soil is certainly not "dusty."

Argentina's premier wine regions are located on alluvial fans. In Mendoza's Valle de Uco district, winemakers refer to *brazos*, or "arms," with reference to the sharply bounded channels of contrasting sediments. The soils vary over a couple of meters from being thick sands to ribbons of coarse pebbles. In the Stag's Leap district of the Napa Valley, winemakers long recognized a "sweet spot" within the Fay vineyard, located on an alluvial fan, which consistently yielded outstanding grapes. Investigations by two geologists, Jonathan Swinchatt and David Howell, revealed that dramatic but hidden changes in the particle size of the sediment was the explanation. At the exact location of the special site, they found that just below the land surface there's a layer containing boulders over half a meter across, swept down from adjacent hills during some ancient violent storm. It provided just the right drainage conditions for this setting. This example also illustrates that the debris in alluvial fans typically has not traveled far: here the boulders came from the volcanic rhyolite that today composes the nearby hills known as the Stag's Leap Palisades.

River Terraces

To look at a second feature that is especially important for viticulture, we need to revisit the idea of a river profile, mentioned earlier. Suppose an entire land surface were to undergo a rise in elevation, or the seafloor were to lower, owing to the action of the Earth's internal stresses. Either way, the river has to start all over again, so to speak, in trying to develop its profile. As a result, various river features arise; we have already met one while discussing rivers cutting a gorge. Similarly, where an already meandering river has to start vigorously cutting down, it will generate **incised**

meanders. Excellent, spectacular examples are the steep-sided hills alongside the sinuous course of the Mosel between Coblenz and Trier.

But particularly important for vineyards is the place where the river readjusts levels and develops river **terraces**. Numerous vineyards around the world are located on these features, perhaps the most famous examples being the left-bank vineyards of Bordeaux. Terraces also form around lakes as a result of fluctuations in lake levels, such as in the Niagara wine-producing area alongside Lake Ontario and beside Lake Okanagan in British Columbia. Here, terraces are called benches, but this should not be confused with the "benches" of California, discussed earlier as part of alluvial fans. We are talking here about terraces that formed naturally. The famous and spectacular vineyard terracing along rivers such as the Mosel, Danube, and Douro or by Lake Geneva in Lavaux, Switzerland is, of course, human-made.

As a general rule in geology, features having to do with deposition become younger upward. Lower-down deposits come about before any higher deposits, which must have formed later (see the law of superposition in Chapter 11). Terraces, however, are different. Although not exclusively so, different terrace heights reflect falling sea levels. Therefore, the first-formed terraces are left stranded higher on the valley sides, while newer ones are formed at new, lower levels (Figure 8.13). In the Médoc-Graves region, some properties sit on the highest terrace, probably having been formed around a million years ago, whereas lower terraces nearer the river are half that age or less. Terraces formed toward the very end of the Ice Age are lowest of all. Some of these terraces are veneered by very recent alluvial sediments, deposited as sea level rose to modern levels.

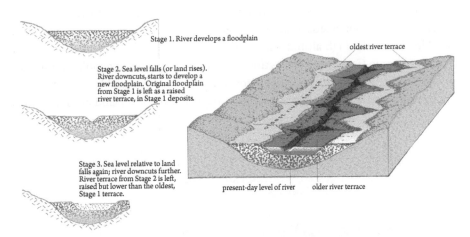

Figure 8.13 Diagrammatic sketch to show the development of river terraces. In this example, formation of an alluvial floodplain (1) is followed by a fall in sea level and a period of river downcutting, to produce a new, lower floodplain (2). Continued downcutting by the river leaves the two abandoned terraces shown in the block diagram.

In line with the variations in floodplain sediments mentioned earlier, because these terraces are remnants of earlier floodplains, the soils within them vary. Even where sands and gravels dominate, subtle differences may have important repercussions. The adjacent chateaux of Lafite-Rothschild and Mouton-Rothschild are located on knolls of slightly differing topographies, but the geology is essentially similar. However, some say that even subtle variations in the grain size of the gravels, and hence drainage, help explain differences between the two wines. Presumably, with this point in mind, winemakers in the areas around the great rivers of Chile—Maipo, Rapel, Maule, and so on—are seeking the most suitable sites on the river terraces there. In other words, both in the Old World and the New, given the right climate, river terraces are well suited to viticulture.

Hillslopes

The bulk of the world's wine may come from river valleys and plains, but in terms of wine quality, many people think of hillslope vineyards. Some of the classic wine styles developed in the more northerly parts of Europe, where sloping land gives advantages in the reduced warmth and sunlight of these higher latitudes. Vineyards on hillslopes catch more of the sun, avoid frost pockets, and are likely to be well drained—all special advantages in moister climates.

The balance between erosion, movement of debris downslope, and possible accumulation, particularly at the foot, determines the overall form of the slope. In a very general way, the top of the slope as it leaves the hilltop is gentle, but it soon steepens to give a convex form. Some hillslopes have a capping cliff of bare rock at the top of the hill (Figure 8.14), with chunks of rock falling away to accumulate as **scree**. Numerous vineyards utilize the good drainage that scree provides, such as in the Alto Adige, the high-altitude sites at the foot of the Andes, in Mendoza, and at Wachtenburg in Germany's Pfalz. Any crags on the slope give it a look reminiscent of a staircase, with the tougher layers forming the stairsteps and less resistant layers forming gentler slopes between. These strips of more level land are sometimes

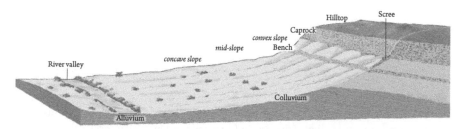

Figure 8.14 Sketch block diagram to show a typical profile of the kind of hillslope, here developed on the scarp slope of a cuesta, that might be used for viticulture.

referred to as **benches**, again as with lake benches, a different usage from that in the Napa and Lake counties of California.

Further downhill, the slope becomes more concave, to an extent depending on the accumulation of downslipped sediment. Where the changeover between the two forms is more than a point, there will be a straight section in the middle of the slope—the **midslope**—which is often favored for viticulture because of its balance between good drainage, exposure to sunshine, and ease of working. The gradient of the midslope depends on the nature of the debris covering it, specifically the frictional resistance it offers to gravity trying to pull it downward. Coarser, more angular sediment may interlock and increase the frictional strength, giving a steeper slope, whereas finer, more clayey slopes will be less strong and the slope angle will be less. The village of Bremm on Germany's Mosel is said to have the steepest vineyards in the world: some say that the jagged, interknitted slate splinters of the Calmont vineyard allow slopes to reach an angle of at least 65 degrees.

The sediment on a hillslope is referred to overall as **colluvium.** It contrasts with alluvium, which is transported by rivers and typically consists of substantial thicknesses of highly mixed detritus brought from diverse sources far away. Colluvium, in contrast, comprises a relatively thin veneer of fragments of the material higher up the same slope. Within the Côte d'Or region, both kinds of sediment occur, but apparently the majority of the Grand Cru vineyards there are located on colluvium.

This colluvium continues to move downhill; in some places, the soil is slipping away faster than it is being replenished. Some estimates have the soil in Burgundy, even where well managed, thinning by about 1mm a year. This doesn't sound like much, but the soil at, say, Aloxe-Corton is less than a meter thick and Burgundy's viticulture is a long-lived enterprise! Elsewhere, in particularly hilly areas, such soil erosion can be a major problem. In Alto Monferrato, for example, in Piemonte, northwest Italy, hillslope soil losses have been measured as exceeding 20 tons per hectare—*each year*. That colluvium can comprise any geologic material is illustrated by Coombsville in the Napa Valley, where it involves the local volcanic bedrock; by central Otago, New Zealand, where it contains various schists and gneisses; and by the slopes flanking the Curicó Valley of central Chile, which are coated by colluvium derived from the granite that forms the hills.

Colluvium will inevitably be on the move downward; it's just a matter of the timescale. Finer material, so-called soil **creep**, may be moving imperceptibly. Its results are visible in trees leaning downslope, bulging walls, and the like. If it gets exceptionally wet, such fine sediment can move much more rapidly, as a **mudflow**. These flows can sometimes move with such energy that they are capable of transporting large boulders for considerable distances in **debris flows**, characteristically wholly unsorted, and having at any point fragments ranging in size from clay to boulders. Examples are common in semiarid regions, but they also arise in humid temperate regions, such as along the eastern flank of the Blue Ridge Mountains, Virginia, and around Terlano, near Bolzano in the Alto Adige.

Coarser material collapsing downslope is a **landslide** (or **landslip**), with the very rapid slip of large rock fragments resulting in a **rockslide**. This is probably the explanation of the jumble of red mud, cobbles, and boulders, much of it composed of granite from nearby higher slopes, in the banks of Jacob's Creek at Rowland Flat in the Barossa Valley. Perhaps there is no more striking vineyard-related example than the great rockslide on which Abymes, just south of Chambéry in Savoie, is founded. Its catastrophic movement on one November night in 1248 killed more than 5000 souls. It reminds us that the kinds of processes we have been discussing are dynamic and that the land surface is continually changing.

Further Reading

Harvey, Adrian. *Introducing Geomorphology: A Guide to Landforms and Processes.* Tampa, FL: Dunedin Press, 2012.
   This is a slightly technical work, though it is not intended as a textbook.

Swinchatt, Jonathan, and Howell, David. *The Winemaker's Dance.* Berkeley: University of California Press, 2004.
   In this very readable book, Swinchatt and Howell report their work on alluvial fans in the Napa Valley, together with other thoughts on the interplay between geology and vineyards.

Australian landforms and geology are described in Chapter 5 of the free download at http://press. anu.edu.au/titles/shaping-a-nation/, with beautiful maps and photographs, though there is little explicit material on viticulture.

9

Weathering, Soil, and the Minerals in Wine

Rock Weathering, or Where Does Soil Come From?

Weathering of rocks is the crucial first step in making vineyards possible. For where the debris produced by weathering—the sediment we met in Chapter 5—becomes mixed with moist humus, it will be capable of supporting higher plant life. And thus we have soil, that fundamental prerequisite of all vineyards, indeed of the world's agriculture. So how does this essential process of weathering come about?

Any bare rock at the Earth's surface is continually under attack. Be it a rocky cliff, a stone cathedral, or a tombstone, there will always be **chemical weathering**— chemical reactions between its surface and the atmosphere A freshly hewn block of building stone may look indestructible, but before long it will start to look a bit discolored and its surface a little crumbly. We are all familiar with an analogy of this: a fresh surface of iron or steel reacting with moisture and oxygen in the air to form the coating we call rust. In his "*Guide to the Lakes*" of England, William Wordsworth put the effects of weathering far more picturesquely: "elementary particles crumbling down, over-spread with an intermixture of colors, like the compound hues of a dove's neck."

A weathered rock is one that is being weakened, broken down. The rock fragments themselves are further attacked, which is why stones in a vineyard often show an outer coating of discolored material, sometimes referred to as a **weathering rind** (Figure 9.1; see Plate 22). If the stone is broken open, it may show multiple zones of differing colors paralleling the outer surface of the fragment and enclosing a core of fresh rock. Iron minerals soon weather to a powdery combination of hematite, goethite, and limonite, and the rock takes on a reddish-brown, rusty-looking color. The great example of such weathering in viticulture is the celebrated terra rossa, but the rosy soils in parts of Western Australia and places further east such as McLaren Vale and the Barossa Valley are also due to iron minerals. Several Australian wines take their names from this "ironstone." It formed in the geological past as well, and around Barossa early settlers used this ancient, tough crust, now largely buried, for some of the region's iconic buildings, including wineries.

Figure 9.1 Chemical weathering giving a discolored brown surface, largely due to iron oxides and hydroxides, to white quartzite pebbles, the famous *galets roulés* of Chateauneuf-du-Pape, France.

Calcareous rocks are particularly susceptible to acids. Even rainwater, because it's normally slightly acidic through dissolving some of the carbon dioxide in the atmosphere to produce weak carbonic acid, over time corrodes and dissolves calcareous materials. This is why although limestone breaks to give jagged edges, in nature they soon become rounded. "Because it dissolves in water" formed the outset of W. H. Auden's acclaimed poem "*In Praise of Limestone*," with its allegories of caves, springs, and "rounded slopes." It's why limestone has a characteristically fluted look with pitted surfaces, both on outcrops and buildings made of limestone, such as the old church in Saint-Emilion (Figure 9.2).

Silicate minerals also respond to weakly acidified water but, in general, much more slowly, producing various clay minerals. The degree of tetrahedral linking of silicate minerals that we saw in Chapter 3 is also the basis of how resistant the minerals are to chemical weathering (Figure 9.3). Increasing linking tends to promote resilience; hence, minerals such as olivine and pyroxene react well before framework silicates like feldspar and quartz. Thus, basalt, with its mafic minerals that have only partially linked tetrahedra, weathers around ten times more quickly than granite, yielding the smectite-rich, fertile soils typical of volcanic areas. Granite, however, is more robust, and though its feldspar weathers to clay minerals, the inert quartz tends to give relatively sterile, sandy soils.

Figure 9.2 Chemical weathering of limestone (*Calcaires à Astéries,* or starfish limestone) in the wall of the church at Saint-Emilion, France, the rock that dominates the Saint-Emilion Plateau and some other parts of Bordeaux.

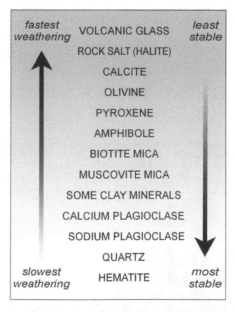

Figure 9.3 Relative rates of weathering of minerals common in vineyards. These are generalizations: the exact order depends on local conditions.

And all the while, rocks are also subjected to **physical weathering**. There's continual abrasion by wind and water, and the daily expansion of bedrock exposed to the sun is followed by contraction at night, perhaps accompanied by ice growing in crevices. The processes all help to disintegrate solid rock, and in some wine regions they produce distinctive features. We saw in Chapter 7 how the exhuming of buried rocks leads to the formation of joints, and relevant here is the tendency for some of those fractures to form within the rock parallel to its external surface. Over time, weathering causes sheets to break away along these joints, in a process known as **exfoliation**. Igneous rocks often weather in this way. Bodies of granite, typically with slightly curving upper boundaries, can exfoliate spectacularly to give rounded **exfoliation domes**. In Australia, examples occur in the Macedon Ranges of central Victoria, the Granite Belt of Queensland, and the Porongurup Range of the Great Southern wine region: the prominent Castle Rock gives its name to a local winery. Many of the wineries around Fredericksburg, in the Hill Country AVA of Texas, are in sight of the Enchanted Rock, another fine example. Paarl rock in the Western Cape, South Africa, shows three exfoliation domes, exposed surfaces of a granite mass that is continuous below the ground.

In basalt, exfoliation typically takes place between sets of joints and is aided by water penetrating along them. The curving rinds of rock spall off, rather like peeling an onion, to give a striking effect known as **spheroidal weathering** (Figure 9.4). There are excellent examples near the vineyards in the extreme east of the island of Madeira, near Machico. In the Yamhill-Carlton AVA, Oregon, outcrops of both basalt and sandstone show the feature, and broken fragments in some of the vineyard soils themselves show spheroidal weathering on a miniature scale. The remnant centers of spheroidally weathered bedrock and of fresh rock between two weathering joints are called **corestones**. With continuing weathering, they will

Figure 9.4 Spheroidal weathering (a) in basalt lava and (b) in sandstone. (Note that the horizontal traces of fine bedding are just visible.)

eventually break loose and may even become part of a vineyard soil. They are well known around Oakville in the Napa Valley, lending their name to a wine there.

Life itself helps degrade rock into soil. We can imagine the effects in a gravel driveway. Microbes and organic debris are always settling on the surface, and specialized organisms such as lichens and mosses may start to grow directly on the stones. On dying, these biological materials will decay to form tiny amounts of humus, with some of it getting trapped between grains just below the surface. Further organic material is always drifting in. Sooner or later, tough little seedlings sprout from the humus and grow up through the gravel. In turn, they will die and decompose, thus adding to the accumulation of organic matter just below the water-washed surface of the driveway. In other words, a soil is forming. Charles Darwin referred to all this as "the transforming power of the minute," but all organisms, big and small, play their part. A recent study showed that the weathering of plagioclase feldspar and olivine was accelerated by the presence of ants. Of course, the contribution of each ant is minute, but then the world has something like ten thousand billion ants!

Soil: What Is It?

Humankind has a long and visceral link with this special material we call soil because of our reliance on it for yielding food—and wine. Indeed, the Canadian soil scientist J. A. Toogood asserted that "soil is the reservoir of life"; Thomas Jefferson (1743-1826) believed that "civilisation itself rests upon the soil". It's a romantic material too, but taking a hard-headed scientific view, we know that for grapevines soil fulfills three functions. One, it's providing a fixed home, a medium in which the roots can extend and be stably anchored. Two, it's the repository of that commodity so vital to vines: water. Three, it's helping to provide the nutrition essential to growth. There is little more to say about the first, and I tackle the second in the next chapter. It's the matter of nutrition that chiefly concerns us here, but first let's discuss a few aspects of the physical nature of this extraordinary material.

The geologic detritus of weathering, the granules of rocks and minerals, forms the soil framework (Figure 9.5). The shape of the particles and how they are fitted together is called the soil **structure**. This governs the spaces between the grains, called **pores**, and the percentage of the soil they occupy, the soil **porosity**. Most of the pores in a soil are filled with some combination of gas and water. The important gas is air, or rather the oxygen that's needed for respiration by the vine roots and the organisms in the soil. It gives rise to the aphorism that grapevines don't like wet feet; it's not so much the water that is detrimental but its exclusion of the vital oxygen.

The average size of its constituent particles is called the soil **texture**, for which we use the terms shown in Figure 9.6. We can roughly gauge it out in a vineyard: a handful of coarse sand will feel distinctly gritty and leave no smears on your skin, whereas loam will leave traces from the clay component. Clay loam can be rolled

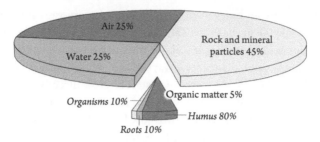

Figure 9.5 Typical proportions of the components of soil.

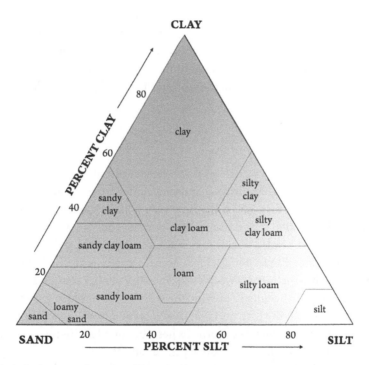

Figure 9.6 Soil texture diagram. Although the conventions are different from the classification of geologic sediment (Figure 5.1), the principles are the same. Thus, each corner of the triangle represents 100% of the labeled component, and the opposite face represents zero percent. The central fields show mixtures in various proportions. Loam is generally considered the ideal soil for most purposes: equal amounts of sand and silt, giving porosity and permeability, and a lesser amount of clay, to help provide nutrients and some cohesion and to retain water.

into little strips, feeling a bit gritty if sand is present; clays and silty clays feel smooth and can form long, pliable ribbons.

Sandy soils aren't able to retain nutrients very well, but some commentators believe they can bring special characteristics to a wine, such as with the *dune sabbiose* (sandy dunes) in Ferrara, Emilia-Romagna (Italy), or the *vins de sables* (wines of the sand) from the Landes dunes, Aquitaine, and the Bouches-du-Rhône, Provence (France). It's unclear why this might be the case, but certainly sandy soils had an historical importance because the dreaded phylloxera louse seems unable to thrive in sandy soils.

Originally called *phylloxera vastatrix*—vastatrix meaning "the devastator"—in the latter half of the nineteenth century phylloxera's ravages in France were becoming almost as drastic as potato blight in Ireland until science came to the rescue. However, some areas with exceptionally sandy soils managed to hold out naturally. Possibly the oldest producing vines in France are those in a tiny vineyard hidden behind a bluff near the village of Sarragachies, in the Gers, southwest France. Even though this is a national shrine, there's no sign by the road so if you're seeking it, look for the rickety white footbridge over a little stream! Planted in 1822—only a year after Napolean Bonaparte died—these monuments survived phylloxera by being rooted in extremely sandy soils (Figure 9.7), part of the huge Lannemezan alluvial fan (Figure 8.12). One of the most historic Australian wineries, Chateau Tahbilk, has vines going back to 1864, through its location on the alluvial sands and gravels of the Goulbourn River which helped deter the phylloxera.

But for vines the heart of the soil is the very stuff that defines it: the organic content. All that rotting vegetation, all those creatures that live in the soil in unimaginable numbers, all secreting organic compounds and then decomposing after death, they all give rise to **humus**. Unlike the soil geology, incompletely formed humus

Figure 9.7 Sandy soil in the Pédebernade vineyard near Sarragachies, Gers, southwest France (Saint-Mont AOC).

has odors, but thankfully, the compounds aren't taken up by the vine roots. Think animal manure! Vines don't need much of this decayed material, though there has to be some inasmuch as it is a vital source of vine nutrition, not least because of its high CEC. Famously, the boulder spreads of Beaucastel in Chateauneuf-du-Pape and the slaty scree of the middle Mosel look hopelessly infertile but below the surface, in the root zone, there's organic matter. And humus is complex stuff. As an indication, just one of its numerous components is humic acid—chemical formula: $C_{185}H_{191}O_{90}N_{10}S$!

A Growing Vine Needs Nutrients

For many centuries, it was believed that vines were made of matter drawn from the soil. Manifestly, the vines were taking water from the ground—if the soil dried out the vines withered—so it was thought that they must be somehow process-ing things in the water to constitute their stems, leaves, and the grapes themselves, including the juice. It seemed obvious, and what alternative was there? A spiritual dimension was also invoked—as by some today—but the idea that the vineyard soil was central to wine seemed self-evident. It all helped enshrine the tradition of a special alliance between soil and wine, which is treasured to this day.

However, a series of scientific advances in the late 1800s revealed that vines don't work like this at all. Scientists began talking about **photosynthesis**, a process super-ficially even more magical. We now know that vines are made not of soil but, in a way, of sunshine, air, and water. Sunlight provides the energy for vine leaves to derive carbon and oxygen from carbon dioxide absorbed from the air, to split water taken in through the roots into hydrogen and oxygen, and to manufacture from these three elements an astonishing number of carbohydrate compounds. And it is these, along with water, that largely constitute vines, grapes, and wine.

So these days there is nothing mystical about how a vine grows in the soil. The soil might conceivably be interacting with the vine in ways not yet known to sci-ence, but the basic functions are clear and demonstrable. Imagine: if we simply plant a vine in freshly ground-up rocks, it will die. It may be in bright sunlight and so potentially it can photosynthesize, but if its roots cannot access water, the vine can-not make the hydrogen needed for its constituent carbohydrate compounds. And even if we moisten the pulverized rock, the vine won't last much longer—rather, it is much like a cut flower in a vase of water. It won't grow. Something else is needed.

Experimentation tells us that the vine's living processes will only work in the presence of certain inorganic elements, iron for instance, though most are only needed in tiny amounts. If we grind up iron filings and spray them on the leaves, the vine will still die, and scattering them on the pulverized rock won't make much difference. Even ramming them down next to the vine roots won't help. Crucially, we have to *dissolve* the iron and the other elements in the water at the roots; then the vine can absorb them and thrive. In fact, it would grow even with its roots just in aerated water containing the dissolved elements, as in the hydroponic system of producing crops. By experimenting with different elements in various trial-and-error amounts, we have established just which elements are essential and in what amounts, exactly as the pioneering German botanist Julius von Sachs did in the latter half of the nineteenth century. Together with sunlight, this is all that a vine needs; unlike organisms like ourselves, they don't have to take in vitamins, proteins, and the rest.

We call the elements essential to growth **nutrients** (Figure 9.8) and sometimes **essential nutrients**. Mycorrhizal fungi living in the soil can extract some of them directly from geologic minerals and transfer them into the vine's roots, but otherwise complex weathering processes and ion exchange have to act on the geology to release the elements into the soil's pore water. So we often refer to these nutrients because, one way and another, they're originating in the vineyard ground, as **mineral nutrients**. Wine writers like to talk about vine roots "seeking minerals and nutrients," but this is tautological. It's different with humans: we need to absorb vitamins, amino acids, fats, and the like, but vines don't. For vines, minerals and nutrients are the same thing. But particularly significant confusion arises where we simply refer to the vine nutrients as **minerals**. In fact, it's something of a flashpoint because the word being used in this way is often misleadingly blurred with the geologic minerals that make the rocks and soils. More of this later.

The weathering processes that unlock and dissolve the constituent elements of geologic minerals operate chiefly at or near the ground surface. By definition, the subsoil and bedrock are little weathered and therefore offer restricted nutrient availability; micro-organisms are similarly sparse down there. So the frequent assertion that deep roots are providing vines with valuable nutrients lacks much basis; most nutrition typically comes from the top few tens of centimeters or so of the soil. Grapevines certainly root more deeply than many other plants, but they are mainly seeking water. And even near the surface, the weathering processes are slow—usually too slow to keep pace with the needs of growing plants. So in nature, at the end of each growing season, vegetation dies back and, on falling to the ground, rots down to help provide humus for next year's growth. In other words, the nutrients are recycled, with continual background "topping up" from the geology.

In agriculture, abstracting crops each year undermines this recycling: as every farmer or gardener knows, to avoid depleting the soil, if you're harvesting crops you have to put something back, perhaps manure or compost, or artificial fertilizer. In vineyards, growers may spread pomace, or they may allow weeds and cover crops to decay. The resulting humus contributes a CEC in an analogous way to clays, as its constituent organic particles are extremely tiny and have numerous surfaces with an overall negative charge. In fact, it has a higher CEC than any clay mineral. Moreover, in a moment we shall see that some of the most critical nutrients are necessarily provided via humus.

Nutrient	Representative functions	Optimum concentration in soil (ppm)	Main sources
Nitrogen	Metabolism of chlorophyll, nucleic acids (hence genetic information); amino acids, hormones, etc.; compounds, e.g., pyrazine, that contribute to wine flavor.	Not normally monitored	Soil bacteria fix ammonia (NH_3) and nitrate (NO_3^-) from the air
Phosphorus	In nucleic acids, cell membranes; involved in cell growth, signaling, energy storage.	20–50	Bacteria release H_2PO_4, HPO_4^{2-} or PO^{3-} from humus
Sulfur	Involved in amino acids, chlorophyll, and proteins, and in plant compounds defending against pathogens.	10–200	Bacteria release SO_4^{2-} from humus
Potassium	Facilitates metabolism, including synthesizing proteins.	75–100	Cation exchange with humus, and with clays derived from K feldspar, muscovite, etc.
Calcium	Involved in enzymes, starch, cellulose, cell walls, and cell signaling.	50–2000	Cation exchange with humus, and with clays derived from Ca feldspar, calcite, etc.
Magnesium	Cofactor for many enzymes. Involved in protein synthesis, ion transport, nucleic acid, cell signaling.	100–250	Cation exchange with humus, and with clays derived from mafic minerals, dolomite, etc.
Iron	Chlorophyll formation; helps activate respiration and photosynthesis.	~20	Fe^{2+} exchange with humus, and with clays derived from mafic minerals, iron oxides, etc.
Zinc	Catalytic, structural, and regulatory functions involving enzymes, proteins, etc.	~2	Zn^{2+} exchange with humus and clays
Manganese	Chlorophyll formation and a range of enzyme functions.	~20	Mn^{2+} exchange with humus and clays
Copper	Proteins and a range of enzyme functions.	~0.5–10	Cu^+ and Cu_2^+ exchange with humus and clays
Molybdenum	Amino acids, enzyme functions.	~0.0004	MoO^{4-} anion exchange with humus and clays
Boron	Carbohydrates, hormones, cell growth.	0.2–12	BO^{3-} anion exchange with humus and clays
Chlorine	Involved with enzymes, osmotic balance, and possibly photosynthesis.	surficial	Airborne Cl^- dissolved via soil surfaces

Figure 9.8 The nutrients essential for healthy vine growth, their target ranges in vineyard soils, and typical sources.

Because grapevines have only modest nutrient requirements, vines can thrive in alarmingly poor-looking soils in which, say, maize, soybeans, or spinach would simply perish. But even the stoniest, most barren-looking vineyard soil has to contain some humus, even if it's out of sight below the rocky soil surface. In other words, although we usually think of nutrients as mineral in origin, in practice much of the nutrition year on year comes not from the geology but from organic material. One could even say that those enthusiasts swirling their wine glasses and declaring a whiff of this or that geological mineral ought to be talking of rotted vegetation and decomposed organisms.

A Look at the Nutrient Minerals

So what are these essential nutrients exactly, and where does the vine get them from? We begin with the nutrient that has the most immediate impact on vine growth and that illustrates the importance of humus—nitrogen. Curiously, although all vines are surrounded by air, which consists largely of nitrogen, they are incapable of tapping into it. And neither are geologic minerals much help as in general they don't contain nitrogen. So the vine has to rely on bacteria that can take in atmospheric nitrogen and turn it into ammonia gas (NH_3), and then other bacteria in the humus that can convert this ammonia to the soluble nitrate (NO_3^-) anion, from which the vine roots can access nitrogen.

It's rather similar with phosphorus, a common *trace* constituent of rocks and soils, particularly as part of the family of calcium phosphate minerals known as apatite (some members of which help form our bones and teeth, and kidney stones). Usually, vines have to access it indirectly through the soil's humus, as soluble phosphate ions. In a few places, geologic processes have concentrated phosphate minerals, mining of which is the main source of phosphorus for fertilizer. In parts of Kentucky and Tennessee, phosphate in the limestone bedrock has allowed a grass species to flourish in the overlying soils that has a striking blue appearance when it runs to seed and that gives strong bones to the horses that eat it. Who would have thought that bedrock geology could influence both the racehorse industry and a genre of music?

The other nutrient that is largely provided by the organic part of the soil is sulfur. It occurs in geologic minerals such as the sulfates and sulfides, but weathering has to unlock it. Pure sulfur, found in some volcanic areas, is insoluble. However, it's readily available from humus through the soluble sulphate ion that vine roots can take up.

The remaining essential nutrients are chiefly metallic elements. Potassium, of course, is the third of the three major components of artificial fertilizer; along with nitrogen and phosphorus it is part of the famous nitrogen-phosphorus-potassium, or NPK, trio. And it's the one vines need in the largest amounts. This would seem to be no problem because it's the seventh most abundant element in the Earth's crust, but as noted later, even in soils rich in feldspar, muscovite, and the like, the amount of potassium that is actually bioavailable may be small. Perhaps we know potassium best through its bitartrate anion, which can precipitate out as the "tartrate" crystals seen as heaps in winery yards or glistening on the bottom of wine corks.

Calcium is widespread in rocks, such as in plagioclase feldspar in igneous rocks and in calcareous sedimentary rocks, and it tends to be made available via cation exchange involving the clay mineral montmorillonite. Only in something like a quartz-gravel might calcium be in short supply. Magnesium is also reasonably common in rocks and is an important constituent of the minerals making mafic rocks such as basalt and gabbro; therefore, it is common in many volcanic areas. Most calcareous rocks, especially those containing dolomite, have some magnesium content.

Iron is the second most abundant metal in the Earth's crust (and far and away the most abundant in the Earth as a whole because it composes much of the core) and is very widespread in rocks. It's a rare rock that doesn't have some iron; the common rusty-red tinge of weathered rocks is usually a giveaway for the presence of iron. Surprisingly then, it's fairly common for vines to be unable to take up the amounts of iron they need. The chemical and physical conditions have to be right: for one thing, iron can only be taken up in its soluble ferrous form (Fe^{2+}). In alkaline soils, iron is likely to be in its relatively insoluble ferric (Fe^{3+}) state, hence the well-known chlorosis problem in limestone soils (Figure 9.9; see Plate 23). Conversely, because the ferrous Fe^{2+} form in more acid soils is soluble, it can be leached away. This is why soils in some parts of the Willamette Valley, Oregon, are quite poor in vine-available iron, despite being derived from a bedrock of iron-rich basalt.

There remain six nutrients to consider which are required in amounts of only a few ppm or less. Zinc is generally only a trace presence in rocks, so probably the chief source is humus; hence, zinc deficiencies are more likely to occur in sandy soils that are low in organic matter. Also, zinc is readily soluble in soil pore water and easily flushed away. For example, the region of southeast Washington–southwest Idaho, known as the Palouse, is now home to a number of wineries but was first exploited for wheat. The initial grain crops were huge, but they soon tailed off alarmingly: it turned out that ploughing had exposed the already zinc-poor soils to flushing, and the nutrient had simply washed away. It has been claimed that because of such mobility, the most widespread trace element deficiency in the world's crops is that of zinc.

Manganese generally occurs in nature along with iron, though in much smaller quantities, but it's concentrated in some soils, in the mineral pyrolusite, such as around Moulin-a-Vent, Beaujolais. Some soils, for example, in the Dão region of Portugal, are reported to have elevated levels of manganese through over-use of fungicides, which can contain the metal.

Copper, usually in its oxide form, cuprite, is also widely but sparsely distributed in rocks, though in sufficient amounts. It is much associated with the prevention

Figure 9.9 Yellowing leaves showing nutrient deficiency (chlorosis). The Chateau Aiguilloux vineyard, south of Montseret in Corbières, Languedoc, southern France, is largely sited on sandstones and conglomerates deposited by rivers, but precipitation from localized lakes has produced narrow strips of limestone in places. Here, beneath the rows of vines running across the middle of the photograph (north–south), calcium-rich soils have hindered the uptake of other nutrients, chiefly iron, which has curbed production of the chlorophyll that gives green foliage.

of molds, particularly through the spraying of Bordeaux and Burgundy mixtures, both of which include copper sulfate. Legend has it that a grower in Bordeaux would spray his grapes that were near a road with a blue copper and lime mixture to deter passers-by from picking them. Then supposedly a local scientist noticed that unlike the unsprayed vines nearby, mold wasn't attacking these blue grapes—and so Bordeaux mixture was born. Such concoctions have proved effective fungicides, so much so that large amounts have been used over the years, resulting in large build-ups of copper in some vineyard soils.

Molybdenum can be rich in some granite soils, but vines require such tiny quantities of it that deficiency is uncommon anyway. One complication, though, is the opposing relationship between molybdenum and copper, technically but colorfully referred to as antagonism. For instance, vines can be poisoned by relatively moderate copper levels if the molybdenum intake is too low; conversely, excess molybdenum can cause copper deficiency at otherwise sufficient copper levels.

The remaining two essential nutrients are taken up as anions. Boron occurs in many geologic minerals but in tiny concentrations as an impurity. The vine probably accesses it mainly from humus, via the borate ion (BO_3^-). Boron minerals, such as borax, sodium borate (one of the very few geologic minerals with a slight taste), do exist but only sparsely. They are important commercially, however, with a quarter of the world's production coming from a single borax mine in southern California, near the town called, fittingly enough, Boron.

Finally, we have chlorine. Tiny amounts of chlorine occur in the trace mineral apatite (see the earlier discussion of phosphorus), in some micas and amphiboles in igneous and metamorphic rocks, and in evaporitic sedimentary rocks. More important in practice is chlorine in the air, which travels widely and settles on the soil to become dissolved in the soil pore water, hence becoming available to roots as the chloride anion (Cl^-).

Chlorine is a good illustration of the point that while most of the nutrients we have mentioned are odorless and largely tasteless, in practice our sensory reactions can depend on the form they take. It's familiar in its Cl_2 form as a gas, with the well-known odor associated with household bleach and swimming pools. But we have a quite different response to it as dissolved Cl^-, chloride. There it's odorless, but it can impart a taste to the solution; particularly if sodium is present, we perceive intense saltiness. If the chlorine is incorporated within organic molecules, the effects are again different. Chlorine can be a constituent of compounds that have no taste or smell at all, such as some amino acids and steroids, or it can help form compounds that have very strong aromas—flavonoids and terpenes, for example. Both contribute to the aroma of wine. In addition, flavonoids help give marijuana its characteristic aroma and make pine trees smell piney; terpenes make hops hoppy.

Geologic Minerals, Nutrient Minerals, and Misunderstandings

The word "mineral" is much used by wine writers these days for rocks, soils, vines, grape juice, wine, and even the taste of wine—but as we have just seen, it can mean different things according to the context. We saw in preceding chapters how ions are bonded to form the complex rigid crystal lattices of geologic minerals—the compounds that amalgamate to form rocks and the physical framework of soils. Here the word "mineral" is used in the *geological* sense. But we have just seen how a growing vine needs certain elements, and because they largely originate in the ground, they are often referred to as mineral nutrients or simply as minerals. Here the word is being used in the *nutrient* sense.

To emphasize this distinction, consider that humans need to ingest certain nutrients to maintain bodily health, and, just as with grapevines, we commonly call them minerals. Thus, for example, a breakfast cereal might be "enriched with minerals." This doesn't mean that pyrite or feldspar is mixed in with the cornflakes; rather, the analysis is telling us how much iron, sodium, or zinc the packet contains. If we go to a health food shop and ask for minerals, we will be shown jars filled not with such things as quartz and mica, but tablets enriched in particular elements—phosphorus, calcium, and the like. In other words, *geologic minerals are not the same thing as nutrient minerals.*

This blurring of the two meanings of "mineral" in vineyards leads to considerable confusion in wine writings. A vineyard at Eitelsbach on the Ruhr is said to be exceptional because it has a "soil full of minerals, including iron, phosphorus, and calcite (*sic*)." I see vineyards lauded as being "rich in minerals," "laced with minerals," and superior because they're "blessed with rich minerals." These are seductive phrases, but what do they mean? As we have seen, all rocks and soils are made of (geologic) minerals, not some more than others. There may be a greater *diversity* of geologic minerals, such as in volcanic soils relative to limestone, but without shutting down the pores, one soil can't have *more* minerals than another. So maybe it means rich in nutrient minerals? But that's the same as saying fertile, and it's pretty much axiomatic in viticulture that highly fertile soils are to be avoided, as they lead to high vigor, lower grape quality and thus poor wine. In any case, we have already noted that the greater part of the nutrition comes from the organic matter in the soil. It's almost as though in populist writings this word "mineral" is acquiring some new, transcendental meaning.

Just because a bedrock contains geologic minerals with a certain element, it by no means follows that the element will be available to the vine as a nutrient. The granite bedrock and soils of Lodi, California, are rich in potassium feldspar, but over 90% of the potassium is locked up in the crystal lattice and is wholly unavailable for nutrition. And of the remaining 10% or so, most is caught *inside* the clay minerals that film the weathering feldspar grains and is said to be "slowly available." Any potassium adsorbed on the surface of the clays will be available for swapping with other cations; this is called "exchangeable potassium," although when freed in solution, it easily leaches away. As a result, less than 2% of the total potassium in the Lodi soils is actually bioavailable, and much smaller values have been measured in places elsewhere. In other words, there is a major disconnect between the parent geologic minerals and the nutrient minerals available to the vine.

When weathering processes have finally made nutrients bioavailable, the soil water still has to physically move them through the soil into proximity with the vine roots. This presents further complications of soil texture, structure, and driving pressures. And, incidentally, the disconnect between geologic and nutrient minerals means that when a particular element is eventually absorbed, the vine won't care, so to speak, what its geologic origin was. Magnesium, for example, is essential to vine growth, but it's immaterial to the vine whether it came from olivine, talc, or dolomite, or, for that matter, from humus or a bag of fertilizer. For the vine, magnesium is magnesium.

The amounts of the different nutrient minerals needed by a vine vary enormously. For example, for every unit of molybdenum that is required, the vine needs a quarter of a million units of potassium. Each element is required within an **optimal range** of values (Figure 9.10). This has a span from the lower limits called **adequacy** up

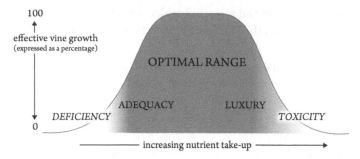

Figure 9.10 Conceptual diagram of amounts of nutrient uptake. The principle holds for all the essential nutrients, but the exact shape and width of the optimal range vary.

to, picturesquely, levels of **luxury**. However, with grape vines the range for most nutrients is surprisingly small. The optimal range of copper, for instance, spans a mere 25 ppm. Lower values lead to a **deficiency** and to consequent growth problems. Lack of just one element will do it, as the vine's metabolism is governed not by the total amount of nutrients available but by the scarcest one. Thus, growers— and other farmers and gardeners for that matter—know that simply adding any old fertilizer doesn't automatically improve performance. This important idea—that a vine's growth depends on its access to the least available nutrient—is sometimes represented by the saying "a chain is only as strong as its weakest link" or by the barrel analogy shown in Figure 9.11.

A common misconception is that the vine roots simply soak up, like blotting paper, whatever happens to be in the soil, as though the roots have to take in whatever the geology throws at them. But the vine's armory of selective devices is intricate and sophisticated. They even help vary the uptake as the growing season progresses, according to fluctuating needs. All organisms require nutrients in differing proportions, but whereas animals like ourselves ingest them in bulk and have internal mechanisms (liver, kidneys, etc.) to sort and expel the excess as waste, plants such as vines regulate them on the way in. How do they do so? Put simply, as water enters the vine roots, it encounters a series of chemical gradients, metabolic mechanisms, and biological "screens" that select which of the dissolved ions can pass through and be transported into the vine system. For instance, there are several kinds of special cell channels that allow the passage of potassium ions while blocking sodium (even though the sodium ion is smaller). In a nice instance of art-meets-science, to mark the 2003 Nobel Prize in Chemistry, one of these very potassium channels in a cell was represented by a steel, wire, and glass sculpture, 1.5 meters tall, and called "Birth of an Idea." It is on display at Rockefeller University in New York City.

Even so, these selectivity mechanisms are far from infallible. The vine must take in soil water in large quantities, and some of the dissolved nutrients will find routes that bypass the screens and be passively absorbed. Consequently, if the soil chemistry is highly imbalanced, say, with unusually high amounts of magnesium or of

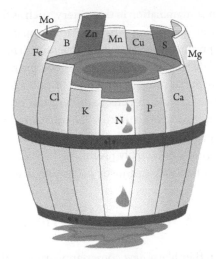

Figure 9.11 A barrel analogy for the concept of a limiting nutrient. Just as the capacity of a barrel is limited by the height of the shortest stave, no matter how long any or all of the others, so the health of a vine is constrained by its access to the least available nutrient. In this analogy, no matter how much various staves are lengthened (i.e., nutrients added to soil), the barrel will not hold more water until the shortest stave, nitrogen in this example, is lengthened.

copper, this imbalance is usually reflected in the chemistry of the grape juice. The soil pH is a major influence on nutrient accessibility (see Figure 10.5). Moreover, the nutrient ions don't act independently but, rather, interfere with each other and even compete for entry according to their various properties. If the ions of different elements are of similar size and properties, just as within minerals as we discussed in Chapter 2, the vine may be, as it were, tricked. All this is why growers are advised to assess nutrition by analyzing parts of the vines themselves, in order to see what the vine is actually taking up rather than what the geology contains. It's also why a good grower patiently walks his vineyard: the key to monitoring its health lies in watching the vines, not in analyzing the soil.

The geologic–nutrient mineral gulf extends as the vine apportions its absorbed elements around its structure. Potassium, for example, is partitioned differently between the woody stems and leaf tissues, and even within a grape: typically, about 37% goes into the skin, 60% into the flesh, and just 3% is taken in by the seeds. The proportions vary from year to year and from season to season. And then there's the matter of the amounts that actually make it through to the finished wine.

The nutrient content of grape juice changes significantly as winemaking proceeds. Fermentation, for example, removes some elements while adding others. Filtering and fining remove ions, and where geologic agents such as bentonite are used, cations may be leached from the materials and added to the must. Additional changes can arise from contamination in the winery, such as from pipes and fermen-tation vessels, and during maturation, when precipitation can occur.

Consequently, the proportions of mineral nutrients in a finished wine bear only a complex, indirect, and distant relationship to the geologic minerals in the vineyard. Incidentally, this is why it has proved so difficult to find a reliable way of using the inorganic constituents of a wine to detect adulteration or to fingerprint its prov-enance. Simple analyses have proved fruitless, so most attempts have involved trace elements, isotopes, ingenious statistics, and so forth. Although most of them typi-cally conclude with various "potentially promising" correlations, in practice, wines subject to counterfeit still rely on diagnostic packaging devices.

A Mineral Taste in Wine?

Early in this century, rather like a new comet appearing in the night sky, the phe-nomenon of minerality in wine arrived. Suddenly wines were described as tasting "mineral," and apparently it has now become the most widely used taste descriptor. This term is even spreading to other commodities: one of the more popular strains of marijuana, Silver Haze, for example, has an "intense mineral nose" (apparently). But unlike other tasting terms for wine, this one is often accompanied by at least an implication of its origin, namely, that it's the taste of minerals in the wine, minerals that were transmitted by the vine from the soil. Sometimes the connection is made explicit, as in "this vine variety, as no other, is capable of transmitting the mineral taste of the bedrock to the wine itself" and in "the vines sip on a cocktail of minerals in the vineyard soil, for us to taste in our wineglass." Some commentators state that these mineral tastes are due to stony soils in the vineyard or roots probing down into bedrock; others report wines with a slate, granite, or limestone minerality. However, all this is at odds with what we have been discussing here.

We have just seen that vines don't absorb geologic minerals, let alone rocks. It is nutrient minerals that are in the wine, but, although wines are described as "mineral crammed," "mineral laden," "brimming with minerals," and the like, their measured concentrations are typically minuscule. Potassium is something of an exception, but even this mineral seldom exceeds much more than 0.15% of the wine; calcium typi-cally ranges between 30 and 120 ppm and magnesium 60–80 ppm, with everything else reported in just a few tens of ppm and less. Occasionally, wines can have higher concentrations, but this is usually due to contamination from agrochemicals, traf-fic pollution, or winery plumbing. Significantly, this can present problems for the winemaker such as stuck fermentations and hazes. In normal wines, mineral nutri-ents typically comprise less than 0.2%, in total.

Can we taste such tiny amounts? Attempts have been made to establish human "detection thresholds" for these inorganic elements but only in drinking water and, although the numbers are subjective and debatable, generally they are higher than the concentrations found in wine. It's possible that the tiny amounts can interact to produce some aggregate effect, but, tellingly, tasters report that as the presence of metal ions becomes increasingly detectable, the water becomes more and more disagreeable. This is hardly a desirable "minerality." And because in water there are few competing flavor compounds, in wine the detection thresholds must be vastly higher (and hence, presumably, even more distasteful).

We can sense some of the organic components known to contribute to wine flavor in truly tiny concentrations: methoxypyrazine, for instance, the compound that gives the green pepper note to some Cabernets, in a few parts per *trillion*. In the presence of hosts of such compounds in wine, the mineral nutrients—though possibly contributing indirectly to our flavor perceptions—are simply swamped. Interestingly, the obvious question of detection thresholds for inorganic elements *in wine* seems unresearched, perhaps because the amounts needed to bring the minerals up to detectable levels could well make the wine toxic.

Judging by their explosive growth, words such as "mineral" and "minerality" are useful for conveying taste sensations, and much has been written about what the terms may actually mean. Scientifically, it remains unclear. But these labels have to be metaphorical and not literal. After all, almost all the other descriptors we use are just that: no one imagines that a wine described as leathery, plummy, jammy, and so on, actually involves these substances. Similarly, describing a wine as mineral or as possessing minerality should not be referring to actual minerals—geologic or nutrient—but should be recalling some cue, some mental association. A recollection may involve some aspect of geologic materials (Chapter 12), but this doesn't mean they are actually there in the wine. All the signs from geology are that "whatever minerality is, it is not the taste of vineyard minerals."

Further Reading

Jackson, Ron. Wine *Science: Principles and Applications* (4th ed.). New York: Academic Press, 2014.
In this magisterial work, Chapters 4 and 5 cover soils and vine nutrition.

Kruckerberg, Arthur. *Geology and Plant Life: The Effects of Landforms and Rock Types on Plants.* Seattle: University of Washington Press, 2004.
This book is fairly technical and isn't about vineyards. I mention it because, unusually, it's about how geology and plant life interact, though it is descriptive and does not consider the processes that might be linking them.

White, Robert E. *Understanding Vineyard Soils* (2nd ed.). New York: Oxford University Press, 2015.
The same author's earlier, more comprehensive book, *Soils for Fine Wines* (New York: Oxford University Press, 2003), is freely available at: https://vinumvine.files.wordpress.com/2011/08/robert-e-white-soils-for-fine-wines.pdf.

A glimpse of the technicalities involved with roots taking up nutrients is at: http://plantcellbiology.masters.grkraj.org/html/Plant_Cellular_Physiology4-Absorption_Of_Mineral_Nutrients.htm

10

Soil, Water, Sunshine, and the Concept of Terroir

What's Beneath a Vineyard?

If we look at a vineyard, it's very tempting to assume that what we see at the surface simply continues on downward. Maybe it does, but most soils vary with depth, and the surface can be quite unrepresentative of down where the work is done, of the materials that surround the vine roots. That's why these days vineyards are peppered with soil pits. Normally, immediately below the surface of the ground is the topsoil, the most fertile part, from which vines get most of their water and nutrients. Below this is increasingly compact, commonly clayey material, **subsoil**, in which relatively little grows. If we continue downward, sooner or later we hit bedrock, for every vineyard sits on bedrock, at some depth or other. Unlike many plants, vine roots can probe many meters downward into the subsoil and even penetrate fissures in the bedrock, particularly if there's a need to seek out supplementary water.

The way soil varies with depth is called its **profile**. The variations in physical and chemical properties may be gradual, or in discrete layers, referred to as soil **horizons**, an arrangement sometimes called a **duplex** soil. A hypothetical example of a layered soil profile is shown in Figure 10.1, and Figure 10.2 gives an example of how a property can vary with depth. The overall depth of a soil above bedrock is termed its **thickness**. In vineyards, this can be anywhere from as little as 20 centimeters, such as at Auxey-Duresses in the Côte d'Or, to alluvium on plains such as California's Central Valley that is measured in hundreds of meters.

Crusts, Pans, and Hardened Layers

Many soil profiles, especially in drier regions, have developed hardened layers. The situation presents us with a real battery of terms: here are the most common. Downward leaching, especially from calcareous soils, can leave a hardened surface crust that is rich in iron minerals, described as being **siderolithic**. *Some wineries in France's Cahors AOC prize their "terroir sidérolithique." Elsewhere in* *western France, in the Atlantique IGP, the iron accumulations were rich enough to be mined, and the landscape is marked by isolated groups of trees growing in the rubble where old diggings have fallen in. Examples occur right across southern Australia and these are usually called* **ironstones**. *Ancient examples are now buried below the ground surface.*

A surface veneer of precipitated calcite is known as a **calcrete**, *and in Spanish-speaking countries and the southwestern United States, as* **caliche**. *It's a term, like ironstone, that is seen on wine labels, in both Spanish and English. If silica is involved, the layers are known, both at surface and where it has precipitated underground, as* **duripans**. *Chemically hardened layers are also sometimes called* **hardpans** *or* **duricrusts**.

Below-ground horizons that have become physically compacted are known as **fragipans,** *which can be important if they pond water above them or they are too tough for roots to penetrate. Fragipans are widespread in the eastern half of the United States. For example, parts of the Ozark Mountain AVA in southern Missouri and northwestern Arkansas have well-developed fragipans that need deep ripping in order to allow roots to grow through them. There are sites in the Pennsylvanian Appalachians up to the Lake Erie AVA that have several meters of soil overlying the bedrock, but because of a strong fragipan horizon, without treatment the vine roots can penetrate down no more than a meter.*

Even where bedrock has weathered in place to yield the overlying soil, its effects (Figure 10.3) can only be very generalized, because of all the permutations of climate, landform, biology, history, and so on, that influence soil profiles. Granite, with its coarse grains and high content of feldspar and quartz, both of which are fairly stable minerals physically, tends to yield sandy, well-drained soils. They are often pale colored, like the parent rock, though in places with a higher manganese content, such as parts of Barolo and Beaujolais, they can have a bluish tone. Progressive weathering of the feldspars, together with the mica also commonly found in granite, produces clay minerals such as kaolinite, but its rather low CEC leads to a nutrient supply that often is little more than adequate. That is, granite soils tend to have low fertility. Although quartz underpins the sandiness, and therefore the drainage, it provides no nutrients. There's no carbonate in granite, and so these soils are almost always acid.

Mafic igneous rocks, in contrast, yield high CEC clays such as montmorillonite. Hence, they tend to be fertile, especially in areas of porous, recently erupted volcanic rocks that weather rapidly. Surprisingly large thicknesses of weathered volcanic deposits can accumulate: some Santorini vineyards have soils over 50 meters thick. Ultramafic rocks such as serpentinite give rise to soils with an exceptionally high magnesium content, which can lead to the nutrient imbalances considered in Chapter 6.

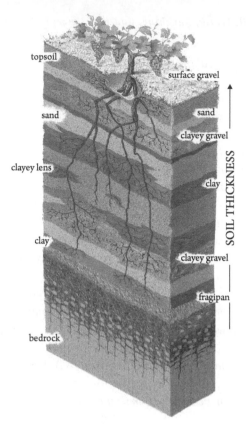

Figure 10.1 Example of a duplex soil profile (imaginary) showing some typical features. The vine roots are able to penetrate the various horizons as far as the fragipan, but they develop lateral roots in the clay horizons and lenses.

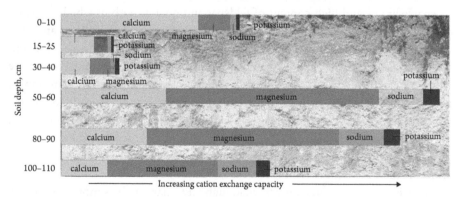

Figure 10.2 Variability of cation exchange capacity with depth (not drawn to scale), here in a kaolinitic soil derived from granite bedrock. The relative availability of selected nutrients is shown; their sum at a particular depth represents the cation exchange capacity for that soil horizon. The surface 10 centimeters has a reasonable CEC, especially for calcium, because of its high organic content, but is low at 15–25 centimeters depth. Three horizons in the subsoil are clay rich—61%, 51%, and 34%, respectively—which give relatively high CEC value, especially for magnesium. From data available at http://www. soilquality.org.au/factsheets/cation-exchange-capacity.

Rock type	Typical soils
Sandstone	*Well drained, inert, poorly water holding*
Shale	*Low to moderate fertility, acidic, can weather to give clayey, poorly permeable soils: surface ponding*
Limestone	*Well-drained, poorly fertile if pure but the alkaline soils give good nutrient access from marly beds*
Basalt	*High potential fertility, acid, can be well drained*
Granite	*Low fertility, sandy soils, well drained but poorly water holding*
Slate/Schist	*Low to moderate fertility, acidic, well-drained when fresh, but fragments weather to give clayey soils*
Gneiss	*Low fertility, acidic, and poorly water holding*
Alluvial deposits	*Rapid variations between poorly drained but fertile clay patches and well-drained, inert sands and gravels*

Figure 10.3 Generalizations about soil characteristics on different types of bedrock.

Soils formed from sedimentary rocks depend heavily on the grain size of the parent for both their drainage and nutritional characteristics. Thus, mudstones and shales yield clayey, poorly drained, and usually rather acid soils but typically with a high CEC. Those from sandstones and conglomerates have better drainage but low values of CEC. Calcareous soils are usually pale, even white, in color, with nutrient and drainage characteristics that depend on the amount of clay present. They are always alkaline to a greater or lesser degree, with all that entails for vine nutrition.

We should remember, though, that while subsoil is commonly derived from the bedrock directly beneath it, the upper soil layers have almost always been moved to some degree. The material may have been transported a great distance, as with alluvial or glacial soils, such that it's now a mixture wholly unrelated to the bedrock on which it now rests. On hillslopes, the soil–bedrock connection is usually much closer, but the soil commonly contains at least some material that has fallen from higher up the slope, which may be quite different in nature. The Kimmeridgian-age (Chapter 11) bedrock underlying the mid-slopes of the Grand Cru vineyards of Chablis, France, is much vaunted as yielding soils superior to those on the Portlandian-age bedrock at the top of the slope. But in fact a substantial amount of Portlandian debris is mixed in with the mid-slope soils, having fallen down the slope.

From Springs to Quicksand: Water in the Ground

Imagine a vine growing in a suitable climate with the nutrients it needs. What determines if it will produce grapes for quality wine? The main geological factor is the water supply: too much water will lead to poor-quality grapes; too little water and the vine will wither. But what is it in the ground that governs the natural water supply to the vine? Two factors are determinative. First, although a small amount of water may be fixed inside the crystal lattice of minerals, particularly in clays, most is in the soil pores. **Saturation** is where all the pores are filled with water: the soil is waterlogged. In practice, the amount of stored water—the **moisture content** or **water content**—depends on the balance between pore size and the clay content, which curbs it draining away. Soils with decent-sized pores but with some clay content store water best.

Second, the water has to get into the pores in the first place and, to be of any use, has to be able to move around. Entry of water into the soil is called **infiltration,** and we loosely refer to onward movement as **percolation** or **drainage**. It is expressed quantitavely by **permeability**. The idea was first worked out by Henri Darcy while he was developing a modern public water supply for the city of Dijon (twenty years before Paris had one). Darcy carried out a series of elegantly simple laboratory experiments that involved feeding water through tubes of sand, in varying situations and conditions. He determined that there was always some intrinsic property of the sand coming to bear, and he called it *perméabilité*. It is now a familiar concept; for example, the rate at which the water trickles through a filter coffee machine depends on the permeability of the ground beans.

So, it's the interplay between the soil's permeability and water content that determines the water supply to the vine and the form of the vine roots. Through their myriad fine hairs, roots can sense the nearby presence of water and adapt their growth to actively seek it out, especially where supply is short. It's illustrated by the contrast between the lateral spread of delicate roots in the thin clayey soils of Pomerol, France, and the deeply probing, robust roots of old Zinfandel vines in the well-drained granite soils of Amador County, California.

Unless the soil is unusually thick, the water characteristics of the underlying bedrock will also be relevant. Sedimentary rocks can have significant porosity if there is not too much cement between grains, and the pores may well be interlinked, giving good permeability. Clayey sedimentary rocks have their mica and clay minerals bonded closely together, such that there is very little porosity between the constituent minerals; most igneous and metamorphic rocks, with their efficiently packed minerals, tend to be similar. At the same time, they often have numerous microcracks within them, and if they are at all interconnected, then water will be able to permeate the rock. We call this **fracture permeability**.

Many limestones show an excellent fracture permeability, which may well partly explain the reverence of some commentators for limestone. The good permeability of the bedrock limestones in the Côte d'Or, France, and Coonawarra, Australia, is largely due to the fractures in the rock. Certainly, good drainage capability is useful in the damp, northerly climates of Europe, particularly if the overlying soils are clayey and poorly drained. Even so, in damper regions, the fracture permeability may not be enough. In parts of Champagne's Marne Valley, the inability of the subsurface limestone to drain percolating water adequately is leading to landslips and extensive damage to some of the vineyards.

Rocks and sediments can store underground water referred to as **groundwater**, and where they are permeable and have good porosities, with adjacent low-permeability materials to curb the water draining away, we have an **aquifer**. These are critically important features, for not only does much agricultural irrigation, including that for vineyards, depend on aquifers, but they are by far the greatest source of the world's drinking water. A wine from Swartland, South Africa, is simply called "Aquifer". Then there are **springs**, a word found in many a winery name and on numerous wine labels. Where downward-percolating groundwater meets an impermeable layer, it may start to move sideways instead, perhaps along an aquifer, and may somewhere meet the ground surface to emerge as a spring. There are all sorts of configurations in which this can happen, many involving variously tilted or faulted strata, and, of course, springs are fairly common in vineyards. Figure 10.4 illustrates the situation around the town of Saint-Emilion.

It all seems so straightforward now, obvious even, but there was a time when springs absolutely beguiled people. Where could that water be coming from? Nobody knew, and springs became imbued with all sorts of mystical, magical, and miraculous qualities. It was Pierre Perrault who in 1674 drew back the veil in Paris with his treatise entitled "*The Origin of Springs*" (*De l'origine des fontaines*). Perrault's rational demonstrations suddenly illuminated millennia of superstition, and his book made a classic contribution to the Enlightenment. (In its own way, it

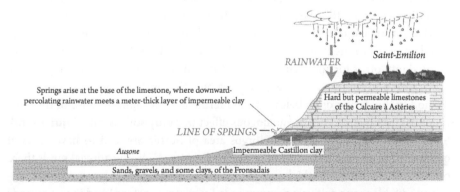

Figure 10.4 Diagrammatic cross section across the southern part of the Saint-Emilion Plateau showing how the spring line at its foot arises where downward-percolating rainwater meets impermeable clay and has to move sideways.

was just as influential as his brother Charles's book *Mother Goose*, which included beloved tales such as Sleeping Beauty, Little Red Riding Hood, Puss in Boots, and Cinderella.)

Water content closely interacts with the strength of a soil (Chapter 7). Soil strength governs such things as whether machines will leave wheel ruts or whether the land can be sculpted, say, to form drainage ditches or terraces. We often refer to wet, clayey soils as being "heavy," in contrast to "light" sandy soils, but this description has nothing to do with weight (actually, clays weigh less than sands, per unit volume). The terms are a legacy of the days of horsepower which determined the work needed to haul machinery.

In general, strength increases with soil depth, and permeability falls. This is because the weight of the soil above—or vineyard machinery—is pushing the particles closer together, reducing the porosity to give an overall decrease in volume properly called **compaction**. It's analogous to squeezing a lump of wet potter's clay between your hands. At first, it's easy to squeeze out a bit of the moisture. The material then gets that little bit stiffer, because the tiny clay particles are being brought closer together and more efficiently packed. Some soils at Pupillin in the Arbois district of the Jura, France, were once so compacted and strengthened by the frequent use of heavy machinery that storm water was wholly unable to penetrate downward. The result was surface scouring and erosion of gullies between the vine rows, which made the sites almost unworkable until vineyard practices were changed.

This interaction of soil water with strength can lead to dramatic consequences. Suppose a clayey soil is being compacted but the permeability is inadequate, such that the water cannot be expelled quickly enough. The water in the pores suddenly finds itself bearing the compacting load instead of the grain framework. And water, being a liquid, has no strength. So the soil finds itself, temporarily, without any strength, behaving like a liquid and unable to bear any force. We call this state **liquefaction**. It can come about through the rapid application of some local compacting load, so a heavy tractor might temporarily sink into the soil. But more widely it's a response to earthquake waves, such that entire buildings may founder. Areas in Marlborough and in Hawkes Bay, New Zealand, for example, have been designated as sites of potential liquefaction. The possibility of liquefaction was incorporated into plans for an extension to the Concannon winery in Livermore Valley, California; and the Kistler winery in Freestone, Sonoma County, California, had to modify the design of its below-ground fermentation room because of the site's potential for liquefaction. The analogous effect in sandy soils is called **quicksand**. Some vineyards in the Subotica-Horgoš area of Serbia are said to have areas of quicksand, as do parts of California's Ballard Canyon AVA. The alluvial sands there can exceed 200 metres in thickness (a report commissioned for one potential winery on suitable grape cultivars recommended asparagus instead) and in wet periods are prone to becoming quicksand. Tractors can sink down into it: vine growers in these places have an unusual extra challenge.

It's a common misconception that water below the ground exists in underground rivers and lakes but, with the exception of cave systems, this is not the case. Groundwater resides in myriad intergranular pores and cracks, as in an aquifer. And if we abstract this water faster than the rate at which it is being replenished by rain, the supply will dwindle. Is all this obvious? Apparently, this is exactly what is happening in some of the world's agricultural regions, including viticultural areas such as Langhorne Creek, South Australia, and Paso Robles, California.

In contrast to wine, in which the inorganic ions tend to join with organic compounds and the flavor is largely due to the aromatics formed during fermentation, groundwater can have something of a taste because of the inorganic ions it contains. Being dissolved directly from the minerals making the host aquifer, groundwater contains anions such as carbonate, bicarbonate, nitrate, sulphate, and halide—all relatively insignificant in wine—that combine with the cations to make salts. Some of these anions have a distinctive, measurable taste (sodium chloride, table salt, is the obvious example) and can be present in substantial amounts. Unlike the case with vines, there is no selectivity. For example, the ion of an element such as uranium, which is the largest one in nature, is rejected by vine roots because of its size, but it occurs in groundwater and hence in some bottled waters. (Some commercial examples contain more uranium than, say, calcium.) The solubility of some of these salts is low, but then there is plenty of time: in the United Kingdom, the average residence time of groundwater in an aquifer is over a century.

Also, in light of some misconceptions, the water in the ground (or, for that matter, a grapevine) is not affected by the Moon's gravity. The water is too constrained to respond quickly enough to the daily changes in the lunar gravitational field, and it lacks sufficient mass anyway. The often proclaimed idea that just because our oceans conspicuously show tides so must everything else, including grapevines, was shown to be wrong over three centuries ago. An object has to have substantial mass and freedom in order to have a significant response to the Moon's gravity. And only four things in Earth's system have that mass and freedom: the oceans, of course; the biggest lakes; the atmosphere; and the solid Earth itself. Yes, the Earth has tides, potentially enormous because of its mass, but the rocks are resilient on the timescale of hours. Nothing else on the planet has sufficient mass to interact gravitationally with the Moon.

Acid Soils, Alkaline Soils, and the Mysterious pH

Farmers and gardeners often talk about the pH of soils, perhaps mentioning a pH of 3 or a pH of 8. As one authority put it, in viticulture and winemaking "the importance of pH cannot be overstated." But what are these numbers, and where do they come from? It's a matter usually confined to chemistry texts, but the accompanying box outlines things at a very simple level. Essentially, pH is a way of quantifying the strength of **acids**, those watery liquids that taste sour and turn blue litmus paper red. And even back in medieval times, some Arabs knew that these acid liquids were canceled out by the waters in which the burnt ashes of a particular Egyptian seaside plant had been boiled, during soapmaking. The name of the plant was kali, and the ashes from it—*al-qalī*—eventually led to the English word **alkali**. So acids and alkalis counteract each other, and a liquid where they are just in balance is **neutral**. This is what the pH numbers are all about.

pH Unveiled

Chemists now know that the key property of acids is the presence of hydrogen (H^+) ions that are freely available for chemical reactions. Conversely, there are none in alkalis, and OH^- ions take over instead; in a neutral liquid, neither kind of ion is available for reaction, as they are virtually all linked together as H_2O. (So once again we are dealing with ions—it was way back at the beginning of Chapter 1 that I suggested Michael Faraday's discovery had far-reaching consequences for vines.)

In practice, however, even in a neutral liquid such as pure water, there are always a few—a very, very few—free, unlinked H^+ and OH^- ions. Their numbers are equal, so there's still neutrality. But the number is truly tiny: one of each for every hundred million million H_2O molecules. So in decimals, the number is 14 places after the decimal point (.00000000000001). With such enormous numbers of zeros, it's easier to deal with a logarithmic scale, which means that where we have 14 decimal places, we write 10^{14}, though because the number here is less than 1, that is, it's behind the decimal point, we write 10^{-14}.

So in pure, neutral water this tiny number of free OH^- ions is exactly balanced by the H^+ ions; that is, there are 10^{-7} of each, and the pH scale expresses this number. To get the pH, we focus in on the logarithmic number for the concentration of hydrogen ions and disregard the negative sign; thus, a neutral liquid has a pH of 7. Then as H^+ concentrations increase and the solution becomes more acid, there are fewer zeros, so the negative logarithmic numbers go down. Hence, acids have a pH of less than 7. Conversely, as the H^+ concentrations decrease (to be replaced by greater numbers of OH^- ions), the pH increases and we have an alkali. Thus, in pH terms, alkalis have a pH greater than 7. Incidentally, in this logarithmic method, each whole-number division in the scale is ten times different from its neighbor: a pH of 6 is slightly acidic, but a pH of 5 ten times more so, and a pH of 4 is a hundred times more acidic. It's exactly the same as with the Richter scale for the amplitude of earthquakes (Chapter 7).

The pH scale was originally devised in the Carlsberg brewery in Denmark to understand how hydrogen ions affected enzymes in beer, although it's still hotly debated what exactly the little "p" was referring to. (Curiously, in light of its significance for wine, the yeast *Brettanomyces* was first isolated in this same brewery, during experiments on the spoilage of British ales, hence the microbe's latinized name for "British fungus"). The pH scale has certainly proved to be a convenient and widely applicable device, including enabling us to describe with precision the acidity or alkalinity of soils—and wines (though strictly speaking, it's the soil *water* that has the pH, seeing as the free H^+ ions have to be in solution). We should, however, remember that in the wine world the term *acid* is used in ways other than pH: as we saw in Chapter 4, so-called acidity in rocks is related to silica content, and in wine we can also refer to total acidity, titratable acidity, volatile acidity, and the like.

It's commonly said that the most important control on soil pH is the nature of the parent rock, but with the very major exception of calcareous rocks, in practice this is not the case in any simple way. Climatic factors, soil management, land use, and soil organisms, for example, are all involved. Granite is often said to give acid soils, and, as we saw in Chapter 4, granite is often called an acid rock. But that is using the word "acid" in two different senses. Actually, granite often does yield acid soils, but so do most other rocks: notably, sandstone, shale, and schist, as long as they lack a significant carbonate content. Consequently, soils on the acid side of pH7 are widespread, and some are significantly acid. On the slates of the Mosel area in Germany, for example, the soils can show values lower than a pH of 6, and on the weathered schists and gneisses of the southern Virginia wine region, lower than 5. In parts of Victoria, Australia, such as the Point Phillip Bay and Pyrenees areas, some pH values are below 4.

Calcareous rocks are sometimes said to yield alkaline soils because of their calcium, but really it's the carbonate that's the bigger contributor. This is borne out by those silicate rocks with a significant calcium content that yield acid soils (e.g., see Figure 4.2). Essentially, in calcareous soils, some of the H^+ ions in their pore waters are bound in bicarbonate ions, and the reduced concentration of free ions gives an increased pH. Thus, around Pouilly Fussé in the Mâconnais, France, for example, the soils are calcareous and have a pH of just over 7, and in parts of Paso Robles, California, they have pH values approaching 10. The values of pH tend to vary across a district: some soils in the El Dorado AVA east of Sacramento, based on granite, have pH values less than 5, whereas the delta of the Sacramento River has calcareous soils with values well in excess of 8. In New York's Finger Lakes region, soils can vary from acid to alkaline across the same vineyard (pH ranges from 5 to over 8 have been recorded), and at the same time pH can vary with depth, in one instance increasing from pH 5.5 to 7 in only 120 centimeters.

Just why is soil pH so important? First, grapevine cultivars and rootstocks vary in their soil preferences. Some say that particular cultivars do best at certain pH values—Chardonnay in (calcareous) soils around pH 7, Gamay and Riesling in more acid soils around pH 6.5, and so on. Perhaps they are influenced by the conditions in the vines' motherland where they evolved and flourished. Similarly, America's *vitis lambrusca* vines evolved in the highly acid granitic and glacial soils of

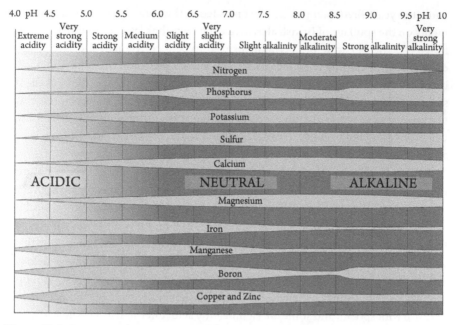

Figure 10.5 Variation of nutrient accessibility with pH. Note that the optimum range of pH is about 6.5–7.5, but with values higher than this, for example, in calcareous soils, the reduced availability of iron can be problematic.

New England, and today they tolerate acid conditions, even below pH 4. Hybrids of the two species, such as Seyval Blanc and Marechal Foch, seem to prefer an intermediate pH range.

Second, microbial and other biological activity is most vigorous in soils between pH 5.5 and 7, which tends to promote a good, open soil texture. Third, soil pH is pivotal for nutrient availability. A recent study of Pinot Noir wines in Oregon found a statistical correlation between a lower soil pH, and thus reduced access to mineral nutrients, and a raised wine pH, within the range to give quality wines. It's well established that ion-exchange capacity and ion solubility, and hence nutrient availability, depend on pH: Figure 10.5 summarizes the situation. This much reproduced diagram forms the basis for most fertilization regimes. Countless manuals describe the symptoms of nutrient deficiency in vines and the appropriate corrective measures—and in all of them pH looms large.

Sunshine Warms the Soil

Thomas Love Peacock (1785–1866, English novelist and doyen of the *East India Company*) remarked that "the juice of the grape is the liquid quintessence of concentrated sunbeams". Modern wine writers often remark that a particular vineyard geology benefits vines by reflecting sunlight to enhance grape ripening and by absorbing the sun's warmth to reradiate it at night. Putting aside for the moment the contradiction between simultaneously reflecting sunshine and absorbing it, we might question how distinctive a factor this is, as judging by populist writings most kinds of rocks—igneous, sedimentary, and metamorphic—can all do it.

For instance, Charles Frankel's 2014 book on French vineyards tells us that the *diorites* at Brouilly "absorb sunlight and later give off the heat to the vines, like a miniature radiator"; that the *limestones* at Sancerre "heat up slowly in the sun, and after dusk radiate their stored-up heat back to the vines"; that at Faugères "one of the great advantages of the *schist* is that it soaks up solar heat during the daytime and radiates it back to the vines after sundown." Elsewhere, I read that "*slate* retains daytime heat to act as a night storage heater" (Jancis Robinson's web-page); that "*granite* is especially heat retentive. In evenings following warm and sunny days it continues to ripen grapes even after the sun has gone down" (Calwineries web-page); that "*gravel* affords excellent heat retention" (J. Patrick Henderson in "All About Wine"); that *flint* "reflects and retains heat" (Charles Fredy in Maui Magazine); that "*basalt* warms up quickly and retains heat" (Stephen Brook in "The Wines of Austria"), and so on (all italics mine). It seems everything does it!

But matters are rather more involved than this. Most of the energy that reaches us from the sun is in the form of bright light, ultraviolet rays, in other words **sunshine**. In a vineyard, the vines will try to capture as much of this light as possible for photosynthesis, but a portion will strike the ground. Most of this light will be absorbed by the soil, but some will be reflected back upward; the proportion reflected is called the soil's **albedo**. Values of albedo theoretically range from 0%, where all the light is absorbed to 100% for a perfectly reflecting surface, but most earth materials fall somewhere in the lower half of this range, depending on their color. Pale colors reflect more and thus have a higher albedo. Quartz sands, being light colored, show values up to 43%, whereas granite sands, with a small amount of darker minerals, average around 35%. Calcareous soils usually fall in the 30–35% range. In contrast, basalt soils average 10%. The moisture content is also an important factor, as wetting tends to darken soils and decrease their albedo, as do humus and iron residues from weathering.

This rather odd word, albedo, has been around for well over two centuries, but it only came into prominence in the 1960s, in the heady days of the Apollo missions aimed at landing a human on the Moon. The choice of a suitable landing site was informed partly by studies of the detailed variations of albedo of the Moon's surface, for the property also depends on texture—the roughness versus smoothness of the surface. A soil's albedo decreases with increasing roughness. So soils rich in clay minerals tend to give smooth surfaces with higher albedos, whereas stones will lower the values (contrary to what is often claimed). Cultivation practices can have an effect too; for example, tillage decreases the albedo, as do weeds and cover crops.

Why does all this matter to vines? One reason is that the amount of sunshine hitting a vine, to which the reflected light may contribute, influences (quite apart from driving photosynthesis) a host of metabolic reactions. For example, anthocyanin development, which is so important in the coloring of red wine, depends on the grape's exposure to sunlight, and the wavelength of reflected light, which depends on the color of the ground, can affect enzyme activity and ultimately the sugar and alcohol content of the wine. But the albedo effect presents complications, such as the angle at which the sunlight is striking the ground. Lower angles give more reflectance, so the effect is more significant at higher latitudes. Thus, some growers in Martinborough, New Zealand, lay white sheets between vine rows to augment the amount of reflected light and enhance ripening. Jerez, Spain, is hardly at a high latitude, but the white *albariza* soils (albedo of 25–30%) can provide helpful reradiation for the palomino grapes grown for dry sherry because they are later maturing than other cultivars.

Then there is the role of bodies of water. Here the albedo effect is also maximized at high latitudes, especially with the glinting effect of choppy water, and at the extremes of the day, when the sunlight is striking at its lowest angle. A local saying has it that the vines around Lake Balaton in Hungary benefit from two suns, one in the sky and one reflecting from the lake. This "lake effect" is also said to help ripen the grapes around Lake Geneva in Switzerland, the Finger Lakes in New York, Lakes Okanagan and Ontario in Canada, and Lake Wanaka in Otago, New Zealand. Vineyards next to the Gironde, Dordogne, and Lot, France, are all said to benefit from proximity to these rivers, as with the Danube in Austria. Apparently, in the south-facing vineyards next to the Mainz River, Germany, in early spring fully 39% of the solar radiation is reflected from the surface of the river.

But there is a further reason why the albedo of a vineyard soil can be relevant: not light but warmth. We have been talking about the portion of sunshine reflected back upward to the vine as light: what about the other part, the energy that is absorbed into the soil? This is the other side of the albedo coin. The less light that is reflected away by a surface, then the more it will warm (it's why we tend not to wear dark-colored clothes in summer). In other words, a vineyard soil cannot both illuminate and warm the vines in similar amounts; the more it does one, the less it affects the other. Growers in the Leithaberg DAC area of Burgenland, Austria, believe that the greater amount of sunlight reflected from their pale limestone soils leads to leaner and more elegant wines, while the darker slates, because of the warmth they retain, yield fuller wines.

For a given albedo, the warming effect on the vines depends on two factors. First, there's the soil's **heat capacity**. Higher values of heat capacity mean that, although the material is slower to warm, it can store more heat and so will be slower to lose it. Actually, the values for different rocks and soils are not great, but basalt, say, has higher values than granite. Hence, the black basalts of Somló Hill, Hungary, give warm soils, and blocks of the bedrock have been used to construct heat-storing walls behind the vines (Figure 8.4). Second is the **conductivity** of the heat, down to where the roots are. With dark soils, solar warmth can affect temperatures down to a meter in the soil, and both root metabolism and nutrient uptake are sensitive to temperature. By some accounts, the Ahr Valley in Germany is the world's most northerly wine region that is dominated by red grapes, whose ripening is considerably helped by the dark soils of slate, basalt, and graywacke. Of course, it only works if the soils are bare. Experiments in Willamette Valley, Oregon, have shown that daytime temperatures above basalt soils drop by almost a fifth if they are covered by grass.

Water has about twice the heat capacity of dry rock but a much lower conductivity, so it makes a great deal of difference whether the soil is dry or wet. This is why some growers irrigate in times of approaching frost: wetting the soil allows it to take in more warmth during the daytime and then release it during the night. Clayey soils have a higher heat capacity than sandy soils, but the main reason they are slower to warm in the spring and cool more slowly than sands is their higher water content. The results of this are well known: hindered flowering for vines on clay soils but an extended ripening season. The same principle applies to bodies of water: Lake Ontario has the smallest surface area of any of the Great Lakes, but because of its depth it stores more warmth than the relatively shallow Lake Erie, and hence produces a much greater effect on temperatures on the Niagara Peninsula that separates the two lakes.

If the vineyard surface heats up, any cooler air above it will be warmed by reradiated heat. The effect is lessened if the air is already warm; in other words, the influence is more significant in cooler regions and at night. But in addition, the amount of reradiated heat falls off very rapidly with distance above the soil, so that only the lowest parts of the vine are affected. Hence, for most trellising systems, which are designed to raise the grape bunches away from the ground, the effect will be small. The reradiation effects of the pale-colored cobbles (*galets*) of some southern Rhône vineyards are much trumpeted. It's hardly a cool climate, but there may be some thermal effect here because the traditional training methods of this region give low-hanging grapes (largely to help avoid damage from the *mistral*), some a mere 20 centimeters from the ground.

Bringing It All Together: Terroir

All the soil properties considered in this chapter, together with the nutrient aspects and the landforms discussed in previous chapters, come together within the concept of terroir, meaning here the natural attributes of a site. It's a decidedly fashionable idea but hardly new—as long ago as 1690, Antoine Furetière in his "*Dictionaire universel*" defined terroir as the "land considered according to its qualities". In France, the concept has not only embraced a cuisine de terroir and a roman (novel) de terroir, but has occasionally soared to improbable heights. Remy de Gourmont, the nineteenth-century Symbolist poet, believed that human characteristics depended on terroir: poets from the igneous and metamorphic rocks of Brittany; cerebral intellects from Jurassic lands; mystics and politicians from Cretaceous areas; and the like. In contrast, today's thinking that grapevines and their fruit are influenced by the site where they grow seems decidedly modest. Exactly how this effect comes about is far from clear, however. A host of interacting factors are potentially involved, and climatic factors are going to be vital. Nevertheless, to many people it's still the land itself and its geology, that's central. So exactly what is the role of geology in terroir?

By far the most important direct contribution of vineyard geology is its influence on soil water. Research in places as diverse as Bordeaux, France; Vaud, Switzerland; Hawkes Bay, New Zealand; the Niagara Peninsula, Canada; and Ningxia, northwest China, has established the connection. Pivotal is the extent to which the geology provides the necessary drainage while allowing an adequate amount of water to be retained, as explained earlier in this chapter. This amount is bound to vary from place to place, though, of course, a similar water status can be achieved by a whole range of different rocks and soils.

Variations in the color of the soil are often said to contribute to terroir differences because of its effect on sunlight and temperature, though as we have just seen, this is likely to be of significance only at higher latitudes and where the grape clusters are trained to be reasonably close to the ground. Similarly, the effect of rivers and lakes is relevant mainly at higher latitudes. A geological role in nutrition is often mentioned, though the modest nutritional needs of grapevines can potentially be met by most kinds of soils, provided they are not overcropped. Also, we have seen that a major contribution comes from the humus content and that the vine is surprisingly selective about its nutrient uptake. However, pH is crucial in governing the availability of nutrients, and that can depend on the geology.

As we saw in Chapter 8, the nature and arrangement of the bedrock and its interplay with erosion fundamentally determine the lay of the land, which in turn produces variations in the local climate, properly called the **mesoclimate**. We may glance at a site and assume that the mesoclimate is uniform, but even slight changes in terrain can lead to crucial differences (as can the position of grape bunches on a vine—the **microclimate**). Where researchers have installed closely spaced sensors to record daily and seasonal fluctuations in such things as air and leaf temperatures, sunlight intensity, humidity, and air flow, the variations have turned out to be surprisingly greater than expected. Local elevation differences turn out to be hugely important. For example, in the Willamette Valley, Oregon, air-temperature fluctuations depend sensitively on precisely how low a site is relative to the surrounding land, what has been called its "relative lowness." Temperature is, of course, a principal driver of vine metabolism, and it can vary much more intricately than a glance might suggest. Measurements of the soil surface in a Pfalz vineyard at midday, for example, vary by 5 °C over just a few meters. Such localized differences have been called the exact opposite of "Segal's law": a man with a watch knows the time, but a man with two watches is never sure. So a grower with one weather station may *think* he knows his vineyard's climate, but a fine network of sensors may reveal a different story.

Then there are the fine variations in microbiology, in both the air and the soils of vineyards. For example, the fungi in the West Auckland, Hawke's Bay, and Marlborough regions of New Zealand differ from one another and are quite distinct from those in Central Otago. Fungal communities associated with Uva di Troia grapes in the San Severo Rosso, Rosso Barletta, Cacc'e Mmitte, and Tavoliere delle Puglie DOC areas of northern Puglia, Italy, are all distinctive. What effect this has for wine is at present unclear, but the kingdom of fungi includes organisms such as *Botrytis,* which is responsible for noble rot in grapes, and yeasts—both those that guide alcoholic fermentation and the so-called spoilage yeasts that can add complexity to wine (e.g., "*Brettanomyces*").

In California, bacteria on Merlot grapes have been shown to vary between individual vineyard blocks, depending on variables such as the precise altitude, humidity, the sun's intensity, and wind speed. The bacteria on Blaufränkisch grapes in Burgenland, Austria, have been shown to produce aromatic compounds that are known to exist in wine, and those that drive malolactic fermentation exist on the skins of Garnacha and Cariñena grapes in Priorat, Spain. These are all fairly new findings, reflecting newly available technologies. So further developments along these lines may be expected: it appears that terroir has to include the microbiology of the site.

Some writers have attempted to rank the various factors that contribute to terroir in order of their importance. To me this is an invidious idea because any hierarchy will vary from site to site and from time to time. I view the known factors as a kind of dynamic, interlinked matrix, constantly changing. Soil drainage, for example, will be more important in some sites than in others, in some seasons than in others, and in some years than in others. And all the factors are interrelated. For example, moistening a soil will increase its heat capacity and so may lead to a warmer soil, thus enhancing the activity of microorganisms, which themselves affect drainage and consume some of the moisture.

Also, there are a couple of wild cards in all of this that must be considered, but which many enthusiasts have turned a blind eye to. First, there's the matter of rootstocks. We usually relate vineyard geology to cultivars, but really it's the rootstocks onto which they have been grafted that interact with the soil. It's the rootstocks that govern water and nutrient uptake, and influence vine vigor. Yet, while wine lovers might be *au fait* with the various Cabernets and Pinots, and even the different clones of Sangiovese and Malbec, to most of us 140 Ruggeri, Kober 5BB, 1616 Couderc, and so on, are an alien world.

Second, it's normal practice around the world to modify the natural factors as necessary, which undermines their importance, and potentially the importance of the terroir as well. Windmills whirl the air on a frosty night; correcting nutrient imbalances is routine. A new vineyard may involve giant machines sculpting the land, and it's *de rigeur* to install drains if the natural drainage is inadequate. Chateau Petrus, for example, which is often cited as a prime example of the importance of terroir, has had artificial drains for centuries to improve the drainage of its clayey soil and has recently augmented them with electric pumps. The steep vineyard slopes alongside parts of the Middle Mosel are remarkable and photogenic, and largely due to river erosion. But not all. The Batterieberg vineyard just south of Enkirch is, as its name might suggest, the result of countless blasts of dynamite.

Dry farming is fashionable where it's possible, but numerous vineyards these days have installed sophisticated irrigation systems that have refashioned the natural regime. And if the water is being brought in from some distance, the chemistry it brings to the vine roots may well be quite different from the indigenous soil water. With regard to the surface color of the soil, for centuries in northerly parts of Europe this has been modified, for instance, by spreading ground-up basalt in parts of the Pfalz (Germany) and by scattering low-grade coals to darken the pale soils of Champagne. Ground cover of one kind or another is commonplace these days. And many hillslope soils require replenishing—sometimes in quantity and sometimes from far afield.

Returning to the natural attributes of a vineyard site, the meaning of terroir being followed here, we have seen that in principle the scientific basis is well understood. But there does seem to be something more to it that science hasn't yet recognized. For example, it would seem that linking the basics of terroir understanding with modern knowledge of viticulture and winemaking should make it possible to produce outstanding wines from almost any site that has a generally suitable climate. You just manipulate the factors and then emulate the practices of some celebrated producer and, presto, you have a rival product. But it doesn't seem to work that way; there is almost a kind of "glass ceiling" that distinguishes a few sites. No doubt the great vineyards of the world have benefited from the money that accrues from their status for maintaining and improving their quality, but it's just possible they have something extra, something yet unacknowledged in science. But perhaps it's more likely that because the interacting factors involved in terroir are fiendishly complex, science hasn't unraveled the very special permutation that happens to come together in each of these exceptional places.

Just as microbiologists are enthusing about being on the brink of new discoveries regarding how vineyard microbiota influence wine, by the same token perhaps one day these special combinations of factors will be elucidated. This prospect of science probing ever further and gradually drawing back veils of mystery makes some wine enthusiasts recoil, but for others it enhances their admiration for the ways of nature and deepens their enjoyment and appreciation of the enduring romance of terroir. As David Mills of University of California Davis said about microbial terroir: "If it demystifies wine, I can understand how some might consider that threatening; I consider it enabling and wonderful."

Further Reading

The reading suggestions in the preceding chapter are also relevant here. In addition,

Fanet, Jacques. *Great Wine Terroirs*. Oakland: University of California Press, 2004.

Parker, Thomas. *Tasting French Terroir: The History of an Idea*. Oakland: University of California Press, 2015.
Parker discusses terroir in its broad, historical sense.

Wilson, James. *Terroir: The Role of Geology, Climate, and Culture in the Making of French Wines*. Oakland: University of California Press, 1999.

11

Vineyards and the Mists
of Geological Time

Geological time is much mentioned in the wine world. Many a label proclaims the geological age of the rocks and soils in which the vines were growing; many a vineyard description enthuses about just how old its bedrock is. The age may be expressed as a fine-sounding technical term or as a quantity, typically, some unimaginably large number of millions of years: "The area's best vineyards are on Turonian soils"; "Cretaceous limestone is best for our vines"; "the wine's secret is the Devonian slate"; "our Shiraz grows in soils 500 million years old." It's almost as though the older the geology can be made to appear, somehow the finer the wine.

I must declare my own position in all this: surely the geological age of the bedrock has little to do with viticulture? The age of the *soil* is certainly relevant, as it is continually changing on a human timescale, but these geological time words almost always are referring to the age of the vineyard bedrock. And almost invariably the age of the soil will be *unrelated and vastly younger than the bedrock*. Surely the vine doesn't care, so to speak, how long ago the bedrock happened to form.

Nevertheless, the fact is that geological time pervades wine literature, so this chapter explains how geologists work with the ages of rocks. The thinking is nicely explained by outlining how geological time was "discovered."

The Dawn of Geology and the Ages of Rocks
Relative Geological Time

Modern geology began two or three centuries ago, essentially when it dawned that answers to questions about the physical world were better answered by going out and observing nature rather than poring over ancient scriptures. We saw in Chapter 1 how James Hutton peered into the "abyss of time." Soon after, other founders of the science began to compare features preserved in rocks with processes they could see happening all around them, and they were able to establish rules (see the accompanying box) that enabled them to disentangle past geological time and to work out the geological

196

history of a particular place. Using these kinds of principles, the early geologists were soon able to recognize past intervals of geological time and give them names.

Some Principles of Working with Geological Time

The procedures the early geologists adopted were given noble titles, although they seem pretty obvious to us now. For example, the principle of superposition says that in a series of sedimentary rock layers, the lowest ones were deposited first and hence are the oldest. Anyone who has poured sand from a bucket to make a pile sees that the deposition begins at the bottom and the pile grows upward. But people didn't do experiments; these were new concepts. So, it became evident that in a series of strata of sedimentary rocks, each layer is younger than the one below it (Figure 11.1).

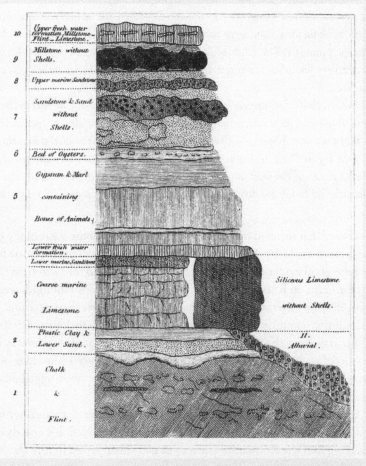

Figure 11.1 Early use of the principle of superposition in determining relative geological time. Early French geologists soon established the sequence of strata around Paris, becoming younger upward from the oldest visible layer, and by 1812 Georges Cuvier was able to include this figure (here in an English translation) reproduced from his classic work on vertebrate fossils: *Recherches sur les ossemens fossiles de quadrupèdes.*

198 VINEYARDS, ROCKS, AND SOILS

The principle of original horizontality observes that all sediments are originally laid down in horizontal layers, so that if we see inclined layers, the tilting must have happened later. The principle of cross-cutting relations says that a particular geologic feature that cuts across another one must be the younger of the two; the other one must have been there already in order to become cross cut. With this kind of armory, the early geologists were in a position to look at, say, some rocky cliff and work out a sequence in which things happened, even if it was in the far distant past.

Less straightforward was linking this sequence with that seen at some other place—what geologists call correlation—but even the pioneers had ways. For example, if a particularly distinctive feature was involved, say an unusual volcanic ash deposit, it could act as a marker. The white, powdery, fossil-rich chalk seen in southern Britain—the foundation of numerous English vineyards—is a very characteristic deposit (Chapter 5). The conditions on the seafloor thereabouts at the time of chalk formation were very special. The hardened and uplifted rock now makes the famous White Cliffs of Dover and the English coast around them, but there are very few places in more distant parts of the world where it occurs. It is distinctive.

However, chalk does appear on the other side of the English Channel in the cliffs around Boulogne, so it seems reasonable to correlate these strata, specifically, to infer that they were physically joined until the Channel was eroded down through them (see Figure 7.2). In France, this same distinctive chalk formation stretches southeastward, all the way into the Champagne region and beyond. So although the occurrences today are dispersed and not continuous, geologists correlate them.

The scheme that resulted from those early studies is still the basis of the geological timescale we use today. Many of the names given to the divisions came from the places where good evidence was first recognized: Devonian for Devon, in England; Permian for Perm, in Russia; Jurassic for the Jura Mountains (though several wine blogs have that the other way round—that the mountains were named after Jurassic time). Cambrian came from Roman attempts to render in Latin, Cambria, the local Celtic name for the land today called Wales: Cymru. The pioneers realized that Cambrian rocks were very old but also that there were even older ones that were exceedingly difficult to disentangle. So they were simply shrugged off as "Pre-Cambrian" (note, for the moment, the hyphen).

This approach whereby we infer that one thing is older than another is called *relative dating*. There are no numerical dates or times involved. It's exactly like historians saying that the Roman invasion of England happened before the Norman Conquest but without giving any dates. This new demystifying of the Earth's history thrilled the public in Victorian times and at the time elevated geology to a heroic science. So it was that at the 1851 Great Exhibition in London, the grounds of the Crystal Palace incorporated a huge representation of the newly revealed geological timescale, complete with oversized models of fossils. In the hollow body of the *iguanodon*, a celebratory fine dinner was staged. ("Sherry, port, moselle and claret" were served.) The geologists of the time could be mightily proud of their achievement.

But there was a problem. Something was glaringly missing. *How old* were these geological periods? *How long* did they last? The Chalk of England and France may be distinctive, but did it all form at exactly the same time? No one knew, and there was no way of finding out. I imagine that those pioneering geologists would have given their right arm for a way to give some numbers for their timescale.

Putting Numbers on the Ages

Geologists call the business of giving an actual age to a rock *dating* it—the source of many an undergraduate quip about geologists loving their subject. The approach is called *numerical dating*. It's like historians, continuing the invading England analogy mentioned above, saying that the Romans first came to England in 55 B.C. and William the Conqueror in 1066. Ingenious ideas were proposed on how this timing could be achieved in geology, such as counting the growth rings on fossil corals, estimating the rate of supposed increasing salinity of the oceans, or determining the time needed for the Earth to have cooled from an inferred primeval fiery state. Unfortunately, there was a suspicion even at the time that the basis of these methods was flawed, and we now know that to be the case.

Another apparent difficulty was that some of these methods gave an age for the Earth in many millions of years. I mentioned back in Chapter 1 how James Hutton "grew giddy" when the immensity of geologic time dawned on him, but millions of years? That even struck some geologists as unthinkably long. For in the centuries preceding all this, various clerics had attempted to work out the age of the Earth from accounts in ancient scriptures, and though estimates varied none of them exceeded 10,000 years. In the English-speaking world, a particularly influential interpretation of various Christian and other ancient texts pinpointed the origin of the Earth as nightfall on a particular October day in 4004 B.C. It became something of a dogma. But even the very first geologists realized that this date gave nowhere near enough time to explain the observed features of the world, although few conceived of many millions of years, let alone the vast intervals known today.

Eventually, bit by bit, a method emerged for dating rocks that was sound and reliable. But it was very slow going. Radioactivity was discovered in the late 1800s, and it took the first decades of the twentieth century to gradually elucidate its nature. Strings of Nobel prizes ensued as it was demonstrated that certain natural elements were intrinsically unstable and through time *decayed*, that is, threw off some of their constituent parts—*radioactivity*—to attain a more stable arrangement.

Understanding of the physics grew ever more sophisticated, and it became clear that the rate of decay of any give radioactive element was fixed—not just roughly constant but an unchanging fundamental property of matter, no more variable than, say, the speed of light or the weight of an electron. Before long, analytical equipment was devised that actually allowed the behavior of short-lived radioactive elements to be observed and tested, in all sorts of conditions, and it all agreed. The rate of decay of any radioactive element was simply unchangeable.

And so it dawned that if we could analyze the amounts of these radioactive elements and their products in rocks then, knowing these rates of decay, we could work out how long the process had been happening, that is, the age of the rock. The radioactive materials, technically called isotopes, are readily distinguished from elements that aren't involved in the decay. Moreover, it could be shown that, for example, a rock solidifying from a magma would lock any radioactive elements into the crystal lattices of its minerals, and decay would immediately begin. So later, the proportions of the as yet undecayed starting element and its distinctive products could be measured and hence the age of the rock calculated.

Theory and laboratory observations agreed: the principle was simple and elegant; the main drawback at the time was the clumsiness of the analytical equipment. And, too, the legacy problem persisted: when this method was tried, the rock ages came out in *hundreds of millions* of years. It was such a mental wrench from what had been believed before. But as the technology improved, as more laboratories around the world came up with the same ages, as the same rocks repeatedly gave the same result and relatively younger or older ones gave corresponding numerical ages, the strictures of earlier thinking just had to be discarded and the method accepted. And for geologists, this finally gave plenty of time to incorporate all the natural processes they observed and worked with, even the excruciatingly slow ones. It all fitted together. So, as the twentieth century progressed, a convincing geological timescale complete with ages became established.

Eventually, further improvements in analytical technology made it possible to date even the oldest rocks on Earth, and geologists began to talk about *billions* (thousand millions) of years. It increasingly looked that the Earth originated well over four billion years ago. I mentioned in Chapter 1 that a good test of scientific understanding is its capacity to predict. So if the consensus was right that all the bodies in our solar system formed at roughly the same time, then if we were to obtain extraterrestrial samples they should give results comparable to Earth's old rocks. They did. At first it was meteorites, thought to be the detritus of a lost planet, then samples from the Moon, and finally (so far) a sample actually measured on Mars. All give ages that agree with an origin a little more than 4.5 billion years ago.

Let's note here that this ~4.5 billion year age for the Earth means that the greater part of its history is occupied by what is now called the **Precambrian**

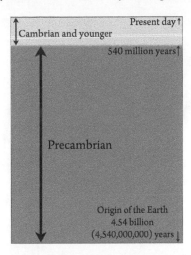

Figure 11.2 Diagram showing the length of Precambrian time in the Earth's history. As it happens, most vineyards are sited on bedrock younger than Precambrian, and so rarely is the timescale shown in this way. Normally, the younger divisions are given at least equal prominence and shown in more detail, as in Figure 11.4.

(note now the lack of hyphen!). It occupies more or less the four billion years of Earth's age, which means the remaining fractional part—the 0.54 billion or 540 million years—covers pretty much all the geological time talked about in wine writings (Figure 11.2). There are vineyards on bedrock a billion years or more in age, such as at Stellenbosch in South Africa, in Western Australia, parts of the Middle Loire in France, and a number of parts of the United States, including Virginia and the Texas Hill Country, but it so happens that most are on geologically young bedrock.

These geological ages are, of course, enormous and beyond our comprehension. As John McPhee put it in his book "Basin and Range": "The human mind may not have evolved enough to be able to comprehend deep time. It may only be able to measure it." But if we are to work with geological time, we have at least to get the number of zeros right. In the old joke, the professor is telling the class that in 5 billion years the Earth will end as it becomes subsumed by the expanding sun, at which point one student was jolted out of his daydreaming. "What's that, Professor, when did you say?" "5 billion years." "Oh, phew! I was worried that you said 5 *million*."

In Croatia today, you can buy a bottle of wine for 5 euros or 7 U.S. dollars; in the 1990s, because of rampant inflation, you'd have to count up the zeros on a banknote such as that in Figure 11.3a (and then realize it wasn't enough). The banknote in Figure 11.3b has even more zeros, a large number in fact, but nevertheless relevant if you were to use it to buy something. Similarly, with geological time we have to get the orders of magnitude right.

Figure 11.3 Reproductions of banknotes showing large numbers of zeroes—daunting
but relevant to purchasing power in the same way as using the orders of magnitude
correctly is necessary for working with geological time. (a) Yugoslav dinaras.
(b) Zimbabwean dollars.

The Geological Timescale

The coming together of the relative and numerical dating approaches has given
us the geological timescale (Figure 11.4; see Plate 24). There are plenty of terminological problems in detail, and a difficulty that arises in the wine world stems
from the plethora of names that the geological pioneers established but that are
no longer used in geology. The early names were often parochial, with no obvious
correspondence to names elsewhere. And so for some decades now, a thoroughly
international geological committee, properly called the International Commission
on Stratigraphy (http://www.stratigraphy.org), has existed, which has a formal and
quasidemocratic system of agreeing on which of the local names should be discarded and which adopted internationally, in the interests of regularizing and har-
monizing usage across the world.

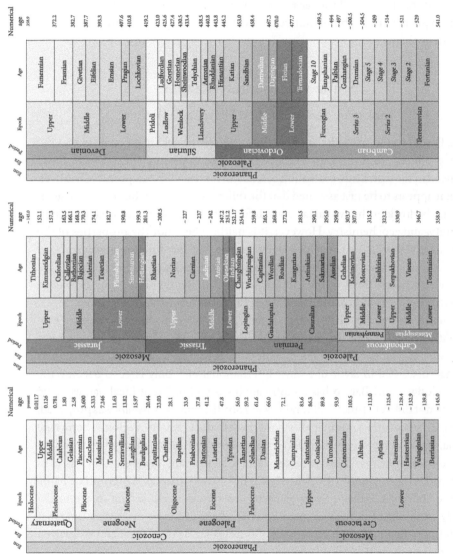

Figure 11.4 The geological timescale (younger than Precambrian) based on the definitive International Commission for Stratigraphy chart, frequently updated at http://www.stratigraphy.org/index.php/ics-chart-timescale.

By and large, the Commission's recommendations have been smoothly absorbed into geology, but, wholly understandably, they have proved slow to trickle into the wine world. The upshot is that a number of terms commonly seen in wine writings have a meaning that now diverges from their correct usage in geology, or, indeed, have become technically obsolete. A few examples of these kinds of things follow. All the terms are seen in wine literature, and all clearly are serving a purpose there, but in different ways they are at variance with the modern usage in geology.

Some Vinous Idiosyncrasies

Tortonian, Serravalian, and Helvetian in Barolo

In the Barolo region of the Italian Piemonte, growers have long recognized that the wines produced from the west of the area tend to differ in character from those produced in the east. Little evaluation of what factors might be responsible has been made: it appears to be just assumed that the difference is due to the changes in geology. The bedrock of the two areas does differ, in such properties as rock type, permeability, and geochemistry. However, it is neither these nor the properties of the (thin) overlying soils that writers usually refer to, but the differing geological *ages* of the bedrock. Thus, we read that the sedimentary rocks of the western area are Tortonian, whereas those in the east are Serravalian. But without prior knowledge these names simply indicate the time when the rocks formed, nothing more.

Whether or not it is this geological age that is relevant, there is another nice illustration here of one of the matters discussed earlier—that of local names disappearing—except that in this case the story is reversed. For the local name was long ago discarded in favor of a more international one, only later to become the official designation! Barolo has an intricate history of different local names being employed at different times, but way back in 1865 the term *Serravalian* was proposed, a name deriving from a municipality right there in the Piemonte region—Serravalle Scrivia. But it never caught on. Helvetian was preferred instead, even though that name obviously derived from relatively distant Switzerland. So Helvetian was used in Piemonte right up until 1957 when it was discovered to be an inappropriate name because Helvetian rocks in Switzerland turned out to be equivalent in age to the already established Burdigalian stage. So "Helvetian" had to be discarded. But what to call the Barolo rocks, which weren't of Burdigalian age? Serravalian was revived and soon was made the official geological term.

Actually, the reference rocks for the Serravalian are now defined in southern Malta, but the point here is that for well over half a century the term *Helvetian* has been obsolete in geology, yet it continues to pervade writings on Barolo and other Piemonte wines. Wine authors, if they are going to highlight the age of the bedrock, ought properly to be using the term *Serravalian* and not Helvetian.

Urgonian

The term *Urgonian* appears in wine writings in confusing ways. It's easy to see why. Urgonian is actually a subtle blending of time, place, overall rock character, fossils, and especially the depositional setting in which certain sedimentary rocks formed. So it's not an interval of geological time: there is no Urgonian on the geological timescale, and it isn't a rock type as such—it doesn't appear in rock classifications. It's reminiscent of Franciscan rocks in California, which some American readers may know of, or the Old Red Sandstone in England. That the term doesn't specify a particular geological time is shown by Urgonian rocks in the Helvetian Alps being well over 130 million years old; those in the Ardèche between 112 and 114 million years; those at Cassis slightly more and those at Tavel and Lirac somewhat less. In other words, there is a span of around 30 million years, involving four geological ages.

In those times, centered on what is now southern Europe but long since vanished, were extensive, warm, shallow waters known by geologists as the Urgonian Sea. The closest analogue today is the Bahamas area, though that is on a lesser scale. The local conditions at the bottom of the Urgonian Sea varied a lot, such that the sedimentary rocks that formed there vary from sandstones to siltstones, from marls to fairly pure limestones. In some places the conditions were inhospitable to life, whereas in others the rocks that formed now display spectacular fossils, nowhere more so than around town of Orgon, just southwest of Avignon, towards the head of the Rhône Delta. The Romans called this place *Urgon*, thus the origin of the term.

One particular rock that formed is an especially strong and massive limestone which today makes a striking presence in the landscape. In Spain, for example, it forms the strange tourist-attracting shapes of the "Enchanted City" near Cuenca. In France, it forms the magnificent cliffs at Cassis and the towering escarpments above vineyards in the Alps and the Jura. The striking *Rocher des Doms*, in the historic heart of Avignon and with panoramic views over the surrounding vineyards (and with the celebrated, now incomplete, bridge at its foot), consists of this richly fossiliferous white limestone, deposited in the Urgonian Sea. Thus, the nature of Urgonian rocks, as well as their geological age, is varied. The term is a tricky one to use correctly!

Muschelkalk

The term *Muschelkalk* has some things in common with Urgonian, being a blurring of time, place, fossils, and rock types. The rock varieties are more limited, however, being largely calcareous and replete with distinctive fossilized seashells, hence the name "shell limestone" (*Muschelkalk* in German).

When relative time was being explored in the early days of geology, it was in Germany that a Triassic period was defined, so named because there it contained three clear divisions. In fact, the term *Muschelkalk* was one of the very first time-related names to appear, being used as far back as 1780. The other two names, Bunter and Keuper, have disappeared from the international timescale but occasionally appear in vineyard writings. Muschelkalk is also officially obsolete in geology but is still often used in Germany and neighboring countries by wine writers and appears on several wine labels.

The setting in which the sediments were deposited in Germany and around was localized, and so it proved difficult to correlate them with rocks elsewhere. Consequently, although the name "Triassic" has been retained internationally, for many years the definitions of its subdivisions have been based on rocks from elsewhere in the world. Thus, the term *Muschelkalk* is used today only in northern and central European vineyards, for the calcareous rocks that formed during middle Triassic times. The sea in which the sediments were deposited became progressively landlocked with time, leading to the limestones and marls in places being rich in dolomite and in gypsum. The shellfish that now characterize these rocks as fossils were adapted to these saline conditions. Muschelkalk underlies parts of southern Pfalz, Rheinhessen and central Franken in Germany, and the Leithaberg, west of Lake Neusiedl in Austria's Burgenland. A number of grand cru vineyards in Alsace broadcast the fact that they are sited on Muschelkalk. Away from this region, the term is not used.

Kimmeridgian

Of all the words alluding to geological time, Kimmeridgian is probably the most familiar to wine enthusiasts. First, let's be clear that although some commentators elevate Kimmeridgian to an almost mythical status, the word is simply the name given to an interval of geological time. Many will think of Kimmeridgian as meaning the marls and limestones that underlie classic wine regions such as Chablis and parts of the Loire, but this is additional knowledge: it is not conveyed by the word "Kimmeridgian" itself. The name simply signifies that the rocks formed in Kimmeridgian times, between 157 and 152 million years ago.

Thus, the rocks at Chablis differ from those around, say, Saumur; the Kimmeridgian rocks at Kimmeridge in England are very different again. About 100 barrels of oil a day are pumped from the black shales there; in fact, much of the North Sea oil originated in organic-rich shales of Kimmeridgian age. Up in northern Scotland, there are Kimmeridgian conglomerates and boulder beds; below ground in northern Holland and in the northern North Sea, up toward Svalbard, the Kimmeridgian rocks are volcanic in nature; in southeast Bulgaria and in northern California and Oregon, some of the Kimmeridgian rocks are metamorphic.

The first person to document rocks of this age was none other than William Smith of geologic map fame, though he called them "Oaktree Clay." Smith first saw them in the grounds of the great country house of Longleat in Wiltshire, but he showed on his map that they continued down to the south coast of England. Later, the rocks there were described by Thomas Webster (1773–1844), the University of London's first Professor of Geology, who referred to them as "Kimmeridge." Webster's term found little acceptance though, not least because his writings discussed how they continued on below the English Channel into France. Why should this be problematic? Because Webster's work coincided with the height of the Napoleonic Wars and pointed anti-French feelings, to the point that literature involving things French was disdained. It's ironic, therefore, that the term was eventually to become established by a Frenchman, through Alcide d'Orbigny (1802–1857) and his *Kimméridgien*. Otherwise, if William Smith's name had survived, we would now be talking about Grand Cru Chablis coming from Oaktreean soils.

Kimmeridgian rocks were originally defined around Ringstead Bay, near the village of Kimmeridge on the famous, fossil-rich "Jurassic Coast" of southern England. For various geological reasons, the Stratigraphy Commission has been deterred from ratifying this historic location as the global reference, as well as all the obvious alternatives, including those in the European wine-growing regions. For nearly a decade now, the Commission has accepted Flodigarry, on the Isle of Skye, Scotland—where the outcrops appear to fulfill all the necessary criteria—as the preferred location for defining the Kimmeridgian, but at the time of writing this location remains unratified.

The Commission's work has meant that some time-honored and much loved names have had to be jettisoned, which of course can touch on nationalistic sensitivities. Famous is the debate in the Chablis area on the extension of *appellation controlée* limits above Kimmeridgian rocks to include the so-called Portlandian. In geology, the term *Portlandian* has for some time been obsolete, following the International Commission's agreement that it should be replaced by *Tithonian* (a name derived from classical Greek mythology). English geologists were much dismayed by this decision, loath to forfeit their beloved name derived from rocks in southern England's Portland (described in Thomas Hardy's *Gibraltar of Wessex*), not least because the limestone there is one of the nation's great building stones. Some of England's most celebrated monuments are constructed from it, including the poignant white headstones of the Commonwealth War Graves, and, astonishingly, 135 buildings in London alone, including St. Paul's Cathedral, Buckingham Palace, the British Museum, and the National Gallery.

In some quarters, such loss of local names has met with chauvinistic outrage. In his book on the wines of Chablis, Bernard Ginestet fulminates against the expulsion of a number of local French names for time intervals in the Jurassic period, proclaiming (in translation from the French original) that it came about because this "geological age . . .was completely monopolized by English men of science." The author goes on to assert this is "why it breaks down into headings which sound like donnish schools of thought." (He is referring to the Kimmeridgian, Oxfordian, and Callovian ages.) "Alas! Another Trafalgar." The beloved French names "went down with all hands in a sea of scientific disdain" and now remain merely "as nostalgic references to the defenders of authentic French culture."

That writer particularly blames a conference held in 1962 in Luxembourg. Although it was attended by "the most eminent geologists from all over the world," at the meeting "one of the bright sparks present (an Anglo-Saxon) proposed to strike off the blackboard with a single stroke . . . the old chronology, and 'Kimmeridgian' was produced like a rabbit from a hat to fill the void." Impassioned indeed. The author does not mention that the term *Jurassic* was introduced by Alexandre Brogniart and that the term *Kimmeridgian* (and, for that matter, *Oxfordian*, *Callovian*, and *Portlandian*) was established by Alcide d'Orbigny, both of whom were French.

Primary Rock

So, as we have seen, a number of time terms are still used in wine writings despite being outdated in geology. But surely the record for this must go to the term *primary rock*. Vineyard descriptions from Germany and Austria, and especially those from the Kremstal and Wachau areas of the Danube, enthusiastically mention the word *Urgestein*, which is usually translated as *primary rock*, or occasionally as *primitive rock*, and dates from the Earth's *primary era*. However, in geology these terms, in both German and English, together with the concepts they imply, have been obsolete for around a couple of centuries!

The geologic materials in those Danube vineyards consist principally of relatively recent loess and alluvium, which contrast with a wide range of older, relatively hard rocks including slate, basalt, granite, gneiss, and schist. These comprise the primary rock. So with prior knowledge, the term can provide a shorthand summary label for that range of rock types, providing a neat contrast with the other group of materials. However, it is used to carry connotations of geological time as well and even implies bringing something profound to a wine.

This is how the term *primary rock* came about. The dawning days of modern geology witnessed the emergence of two rival schools of thought. One emphasized the role of the Earth's internal heat, as an engine for driving processes and creating rocks like basalt and granite which had once been molten (i.e., the thinking that forms the basis of geological understanding today). The other school of thought involved the primordial Earth being entirely covered by a lifeless ocean, from which metamorphic and igneous rocks were *chemically precipitated*. These were the primary rocks, defining the primary era of the planet's history. Then, when life appeared in the ocean, sedimentary rocks formed, containing fossils, and these were secondary rocks, from the secondary era. (Other divisions were added later, but none of them survive today.)

This latter thinking was widely admired and influential at the time, not only in its Saxon heartland but also in Britain and other parts of Europe—but only briefly. For in addition to physical issues of generating and then disposing of such implausible volumes of ocean water, incompatible geological evidence rapidly accumulated. For example, igneous and metamorphic rocks were reported that were demonstrably much younger than fossiliferous rocks; also, metamorphic rocks were shown to be changed variants of rocks that went before. That's why today these rock names have no age connotations.

A Word on Fossils and Wine

Many people are fascinated by fossils, those memories of our planet's distant past. And fossilized seashells catch the eye in a number of the world's vineyards: Chablis and Sancerre in France are well-known examples; the Cederberg Mountains of South Africa and Central Hawkes Bay, New Zealand, are other instances. Some wine writers make much of the presence of these fossils, believing they bring special qualities to the flavor of the resulting wine, and some even maintain that they can be tasted: "a *goût de fossiles* (taste of fossils)".

When we taste wines from soils that we know are rich in marine fossils we may be reminded of things to do with the sea, and it becomes easy to perceive marine elements in the wine. Thus, commentators often mention the abundant and striking oyster-like fossils in the Chablis vineyards (Figure 11.5) and go on to describe the wine as having an iodine taste (Chapter 12). However, as far as the vine is

Figure 11.5 Fossils in the bedrock at Chablis. Dominant are the comma-shaped shells (many seen here in cross section), replicas of the extinct organism *Exogyra virgula*. Wine writings commonly refer to them as oysters, but they are not the same as the oysters we eat today.

concerned, any fossils in the rocks and soils are indistinguishable from any other piece of geologic mineral—for that is what almost all fossils are.

Nearly all fossils come about because the material of the original organism has been wholly replicated by durable geologic minerals, most commonly calcite and quartz. Organisms vanish quickly after death. The tissues are soon scavenged or they rapidly decompose. However, any hard parts such as teeth, shells, or bones survive just a little longer and so have a better chance of becoming fossilized. As a first step, usually the organic remains are rapidly covered by sediment, which isolates them from scavengers and bacterial decay. This is why the majority of fossils are only the hard parts of marine organisms.

Then, as the sediment becomes lithified, internal rearrangement of the organic remnants, or replacement by geologic minerals precipitated from percolating water, produces durable material while preserving the form of the original. A different means of fossilization is for the biological material to be dissolved away, within the lithifying sediment, leaving an imprint in the host rock. This becomes a kind of mold, and with time it becomes in-filled with geologic material, giving a cast of the original shape.

But whichever mechanism was involved, *the fossil is a replica* normally consisting wholly of geologic material. In special circumstances, the original organism can be preserved, such as insects in amber, mammals fallen into tar-pits, and mammoths frozen in ice, but these are exceptional. The overwhelming majority of fossils are facsimiles and are simply geologic minerals making a particular shape. Consequently, fossil material brings nothing different to the nutrition of the vines or the composition of the resulting wine from that of any other geologic mineral.

The Age of Bedrock versus the Age of Soils

In wine writings, the age of the vineyard soils is commonly treated as identical to that of the bedrock, but almost always this treatment will be wrong. For example, descriptions of vineyards located on Precambrian bedrock enthuse about how primeval the soils are; we're told that vines in the Middle Mosel, Germany, are growing in Devonian soils, that the Côtes de Provence, France, has Permian soils. But it's the *bedrock* that has these ages, not the soil. As we saw in Chapter 9, it's the slow breakdown of preexisting rock by weathering that over time produces soil, and it's an ongoing process.

Promotional literature about the Heathcote wine region in Victoria, Australia, is fond of enthusing about the vast antiquity of its Cambrian age soils, but the *soils* are probably a few million years of age at the very most and are still forming. The soils of the Gimblett Gravels of Hawkes Bay, New Zealand, formed in a flood little more than a century ago! Vines may have water-seeking roots probing bedrock fissures, but the bulk of the root system operates in the loose, overlying soil. And for transported soils such as alluvium and loess, the bedrock age is quite irrelevant.

The actual age of a soil is very difficult to establish. The radiometric and other techniques used to date rocks just don't work for soils, not least because they are always evolving. Estimates of the rates at which soil forms from bedrock have been made, but even these estimates are difficult because they depend greatly on climate and a host of other things. But it's clear that while soils are geologically young, by human standards the processes are pretty slow, especially in the absence of moisture. This is why ancient buildings in arid places are relatively well preserved: think of Petra in Jordan or the Library of Celsus in Turkey.

Even for humid, temperate regions, one estimate has it that very roughly a million years are needed to weather just a centimeter down into granite. Another estimates that it takes around 100 years to generate 2 to 5 centimeters of fertile soil from volcanic ash in the warm, humid tropics and 5000 years to produce 1 centimeter of soil on hard rocks in cool temperate climates. One commonly quoted value is 100 years for 1 centimeter of soil development under permanent grasslands in temperate climates. Although these figures vary greatly, overall they mean that soil ages are measured in thousands up to a few millions of years at the most. Obviously, that's a long time to us, but the underlying point here is that such values contrast markedly with the hundred million year ages and more of much bedrock.

Moreover, as the soil develops, it is vulnerable to erosion. For example, whatever soils existed before Pleistocene times in places like northern Europe, Tasmania, and New Zealand were stripped away by the ice sheets of those times, so the soils there today are no more than 12,000 years or so in age. Chapter 9 outlined how viticulture itself continuously alters various chemical and physical properties of soils. For instance, to stretch the point a little, every time a grower drives his tractor or even walks in his vineyard, he will to some extent be compacting the soil and changing some of its physical properties. Every harvest modifies the soil chemistry somewhat.

We can talk very generally about young, mature, and old soils, or perhaps about their stage of nutrient leaching, and indicating something about a soil's age is helpful. Such things are relevant to viticulture, and in practice they are far more valuable than declaring some distant geological age for the parent bedrock.

Further Reading

Web content on the age of the Earth tends to present a frenzy of scientific versus nonscientific positions; authoritative popular books are elusive. Perhaps the most reliable accounts of the discovery of the extent of geological time (and evolution) remain the following:

Dalrymple, G. Brent. *The Age of the Earth*. Stanford, CA: Stanford University Press, 1994.

Gould, Stephen Jay. *Time's Arrow, Time's Cycle: Myth and Metaphor in the Discovery of Geological Time*. Cambridge, MA: Harvard University Press, 1988.

Sections on geological time are included in the kinds of general introductory books listed at the end of Chapter 1.

12

Epilogue

So is Vineyard Geology Important for Wine Taste?

We have seen in previous chapters how grapevines interact with rocks and soils, and in Chapter 10 I discussed the role of geology in terroir. But a question remains, one that is probably uppermost in the mind of many a wine lover: to what extent does geology affect the *taste of the wine in your glass*? I argued in Chapter 9 that the perception of a mineral taste in wine can't have a literal meaning, but what about other tastes ascribed to geology? We might reasonably expect that the geological influences on vine growth have at least some role in wine flavor, but what? Many populist wine writings imply that the answers to such questions are clear-cut, but unfortunately they aren't. Claims are routinely made in wine descriptions that sound fine but that don't easily tally with scientific understanding.

In other words, there's some divergence between popular beliefs and scientific understanding of the geology—wine flavor connection. Part of the explanation may be that many of the populist assertions seem to be based on custom and on anecdote—narratives passed on enthusiastically but *unquestioningly* between wine fans. Two situations are common. First, a description of a wine casually mentions the kind of geology where it originated, implying a significance but without any justification or indication of how it might come about. I give illustrations of this in the following section.

Second, some character of a wine is ascribed to particular rocks and soils but without providing any rationale. For example, a Riesling from Kamptal's Gaisberg vineyard (Austria) is said to have "complexity because of the slaty para-gneiss, amphibolite, and mica" soils. But there's no indication of how these two very specific rock types together with this particular mineral bring this complexity about. Decanter magazine (2016) asserts that in the Elephant Hill area of Hawke's Bay, New Zealand, "free-draining stones give minerality to the wines, while the clay soils produce complex aromatic aromas (*sic*) and flavors." Such bold, unqualified statements suggest that geology–wine flavor connections are well known, well established, and well understood. But they aren't. That clay soils affect drainage and vine performance is clear, but what this means for wine aroma and complexity is not. Stones are inert—otherwise they wouldn't be there as stones—so scientifically it's hard to see how they can imbue a wine with certain flavors.

Elsewhere I read that the limestone of Alsace's Hengst vineyard gives "big, strapping wines"; a Côte-Rôtie wine has a "schist-inspired clarity"; a wine from the Willcox AVA in Arizona is "muscular and masculine" because it came from granite soils. No explanation, no caveats are given: the direct role of the geology seems to be just "taken as read." Austria's Traisental wines produced from limestone vineyards are "less baroque and exotic than those from loess." My point concerns not these descriptions of the wines but the absence of any clue as to how the geology produces these perceived differences. A Chilean wine from the Maipo Valley has "a particular energy because of the granite soils." Putting aside what "energy" might actually mean, how does the commentator know that it's the granite and nothing else that's responsible? How does the granite do it? It's hard to see what the scientific basis might be.

The schism may also be partly due to a cause-and-effect situation. Where wines from neighboring sites that are made in exactly the same way turn out differently, if the soils are different then it's very tempting to pounce on that as the reason— because they're easily visible! You can readily see the surface soil, pick it up, hold it, photograph it, but the other things that will be changing along with the geology, such as the ever-changing array of interacting factors to do with the local climate and microbiology, can't really be seen.

In Chapter 10, I mentioned the importance of "relative lowness" and intricate temperature variations. As a further example of the importance of unseen factors, at the Fault Line vineyards of Abacela winery in the Umpqua Valley, Oregon, the wines coming from each side of the fault line are different in character. And the geology contrasts across the fault, so the usual assumption would be that therefore that's the explanation: it's the geology that's responsible for the differences. No doubt that will be part of the explanation, perhaps through differing drainage properties, but here the vineyard owners have collected detailed temperature data across the adjacent sites and they reveal surprising variations. For instance, on one day the ripening temperatures (actually part of the parameter called growing degree days) varied across the land surface on either side of the fault by as much as 56%. That's potentially a big difference in grape ripening and hence in the character of the finished wines. But details like this require patient data collection and analysis such that the factors tend to get overlooked. No doubt, collating and reporting data on intangible technical details like air velocity, UV intensity, spectral wavelength, and bacterial taxa make for less exciting journalism than what to some is the charisma of geology.

The popular geology–taste image, moreover, fits nicely with the cherished special relationship between wine and the soil; because it's not easily replicated elsewhere, this can be very helpful in marketing. There are a couple of curious aspects to the idea, though. First, this link with the soil isn't being claimed for tasting raw

a crop that actually grows in the ground—a carrot, say. Carrot roots are conspicuously influenced by the physical nature of the soil, as any vegetable gardener knows,
and nutritionally they absorb from the soil pretty much the same things as a grapevine. Yet the taste connection is being applied not to carrots but to a drink, and one
that only emerges after extensive processing of a ripened fruit.

Second, its prevalence is a recent phenomenon. The notion of a link between
soil and wine is ancient, but its present prominence in taste descriptions is less than
a couple of decades old, in the English language at least. Among my collection of
wine books from the later part of the twentieth century, there is not a single one
of them that mentions "geology"; some mention is made of slate in Germany and
gravels in Bordeaux, but otherwise a vague remark on "suitable soil" is as far as they
go. The books I have from elsewhere in the world don't even mention soil. This has
all changed rather quickly and for no clear scientific reason. So, although today it's
fashionable—*de rigeur*, even—to link wine taste with geology, the basis is unclear.

Yet, the scientific understanding is tentative. Most research effort has focused
on soils and vines; not much has extended to wine *taste*. In contrast, the amount
of accumulated anecdote is huge, some of it being based on the working experience of thoughtful practitioners. Moreover, the practicalities of raising grapes and
crafting wine informs in a way that can be much more meaningful than scientific
logic. So is science perhaps missing something? The scientific position is always
evolving because scientists are always asking questions, always probing. By its
very nature, scientific understanding will never be complete. (Saying that science
has *proved* something is always a giveaway for a flawed understanding of how science works. Evidence may be overwhelming, but in science, unlike in mathemat-
ics or law, things are never final.) In other words, scientific understanding could,
and almost certainly will, change. After all, if I had written this book while I was
an undergraduate student, I wouldn't have known about plate tectonics in geology, and not too long before that the role of mineral nutrients wasn't recognized.
Be that as it may, where do we stand just now? Let's look further at the present
situation.

The Label Tells Me the Kind of Soil, So …

I have in front of me a bottle of Cabernet Sauvignon wine from a New World country, and the back label tells me that it was produced "from granite soils." It's typical
of this kind of wine–geology statement. I like seeing such information because it
helps round out the picture and give context to the wine, in the same way that I like
to know about the vine age, the wine's heritage, the winemaker, and the like. But
those details aren't mentioned on this label: the fact that space has been found in
the five lines of terse text to mention the geology seems to imply that this geological
fact is especially important, that it's telling me something about what to expect in

the wine. If it said produced "from sunny slopes" I would infer a fuller, riper style of wine but what might "granite soils" be telling me?

I can't invoke a comparison, because the world's benchmark Cabernets aren't produced from granite soils (e.g., Médocs are from gravels, most Napa examples are from alluvial or volcanic soils, Coonawarra wines are from limestone-based terra rossa), and there are fine and varied examples from all sorts of different rocks and soils. Conversely, granite isn't known to yield wines with particular characteristics. For instance, I read that the granite bedrock at Dambach in Alsace gives wines "beautiful elegance and a very fine fruitiness," whereas those produced on granite at Cornas are described as "impenetrable," "meaty," "powerful," and "brutal." So, the mention on my Cabernet Sauvignon label that the soils are granitic isn't telling me anything more than just that. I wonder if wine reviewers and label writers aren't bound by custom, by a nice sounding phrase, or simply by a longing to nurture the link between the land and wine.

One much cited connection is that between Champagne and chalk. For example, commentators say that chalk brings "lightness and finesse," that it gives the wine its own "particular mineral flavor"; they even talk of the "magic of chalk." Being a kind of limestone, chalk gives soils of suitable pH, it may be soft enough to allow some root penetration, and its good drainage properties are crucial, given the maritime influence on the Champagne region's climate. But how this translates into wine character is unexplained. It's difficult to see what special qualities chalk might bring, especially as it's only characteristic of a part of the Champagne region and has long been manipulated anyway (e.g., carbon in Chapter 2). Of interest are the remarks of d'Omalius d'Halloy, the great early Belgian geologist and pioneer of chalk studies. (In 1822, he wrote of a *Terrain crétacé* that led to the time-name Cretaceous.) He reported that "the soil of the true chalk is not generally favourable to the vine," and he spoke "of the error (when speaking of Champagne) of associating the idea of a chalky soil and a country producing good wines".

If such uncertainties arise with tightly defined rocks such as granite and chalk, then things are even more confusing with those rock names that cover materials that vary greatly in their chemical and physical properties, as we saw in earlier chapters when discussing slate, volcanic rocks, schist, and so on. So to me at least, simply saying a wine comes from a particular rock doesn't convey anything about what I might expect from the wine.

I also have in front of me a table listing the names of the Grand Cru vineyards of Alsace, their particular soils, and the style of wines they yield. The table looks impressive, but it's puzzling. I read that some of the soils are characterized by the elements they contain (iron, magnesium), some by their geologic minerals (gypsum, quartz), and some by geological time (Keuper, Devonian). Other soils are volcanic (tufa (*sic*), ash, "ashes") or are derived from granite or something called "two-mica granite". What should I be drawing from all this information? An array of flavor aspects are listed (floral, discrete elegance, power, minerally, etc.), but nowhere is there any indication of how a connection with the soils might work. There seems to be no correlation anyway.

Generally, the links that are claimed between types of rock and wine character are wildly erratic, but in a couple of cases there is a little more consistency, although even these cases contain plenty of flat-out contradictions. Clayey soils are often said to bring "body" and "structure" to wines, presumably through their characteristic water-holding properties—somehow. Limestone soils are frequently claimed to imbue wines with attributes like "liveliness," "edge," "nervousness," and "finesse." And although limestones are varied in nature, they do consistently give well-drained, high-pH soils. Again presumably, then, it's these properties that are somehow involved. But then we glimpsed above "big, strapping wines" from limestone, and elsewhere I read that Burgundy's "limestone-rich soils give unmatched levels of luscious flavor, one of the world's most opulent." One author tells us that while France's Bourgueil wines from gravelly sites are "light and fruity," those produced from limestone are "full-bodied and well-structured" and "should be given at least two or three years" ageing. It's all a bit unclear. However, some wine enthusiasts hold on to one notion, as follows, that is completely at odds with scientific understanding.

But We Can't Taste Rocks in Wine

"You can really taste the volcanic minerals in the wine." "He speaks of his granitic soil and the wine in your glass tastes of it." "You can taste the Devonian slate right there in the glass." The idea that a vineyard's geology can literally be tasted in your wine is enthusiastically promoted by some devotees, prolonging a belief dating back to times before the discovery of photosynthesis when wine was thought to be made from soil, and today very much chiming with the yearning for a closer connection with the land. Unfortunately, it's scientifically untenable for at least two reasons: (1) vine roots are incapable of absorbing the solid, complex compounds that make the rocks and soil of a vineyard; and (2) geologic materials are generally tasteless and odorless anyway. As the geologist Kevin Pogue put it at the 2017 Oregon Wine Symposium: "Tasting the limestone may sound romantic, but is no different than saying you can taste the 30° slope or the 220° aspect. Limestone doesn't have a flavor and neither does the 220° aspect".

We learned in Chapters 2 and 3 about the complexities of the solid compounds we call geologic minerals, and in Chapter 9 we found that vine roots can only take up nutrient minerals if they're dissolved, as ions in solution. The roots simply cannot take in solid compounds, let alone the rigid aggregates that form rocks. It's not only demonstrable in experiment, but with today's powerful technologies we can actually see the cells in roots and inspect what they are and are not absorbing. And, incidentally, grapevines don't take in anything different than other plants;

they all absorb more or less the same nutrients. Most of what we taste in wine is produced during vinification. This is what primarily sets wines apart, not what the roots absorb. In certain circumstances, substances in the air can land on a vine's foliage and grape skins, and hence directly enter the wine must. They may then go on to influence wine composition and flavor, as demonstrated for vines growing near eucalyptus trees or affected by smoke from forest fires, and just possibly the well-known herby fragrance of *garrigue*. But these are special situations that are by-passing the normal growth mechanisms involving the roots.

Regarding flavor, that is, odor and taste, most of what we perceive depends on smell. The taste components mainly involve ions and compounds in solution, and geologic minerals are practically insoluble. The only significant exceptions are some halide minerals, especially halite (sodium chloride). It rapidly dissolves and, of course, gives the taste of saltiness on the tongue. However, growers avoid salt in vineyard soils, and grapevines try to reject sodium. Hence, wine normally contains little salt, less than the minimum of several hundred ppm that most people require to be present *in water* in order to recognize it: a perception of saltiness in wine is usually metaphorical. The organic compounds that prompt our tasting sweetness, bitterness, umami, and the like just aren't present in geologic minerals.

Odor (aroma or smell) requires that a substance be a vapor in order for us to sense it; thus, for a solid or liquid to be aromatic, it must easily vaporize. Rocks and minerals don't do this. A substance's tendency to volatilize is indicated by a property known as **vapor pressure**: the higher the value, the greater the tendency to vaporize. For many of the organic compounds found in wine—esters, thiols, terpenes—the values are very high, which is why they are collectively called aromatic molecules. Alcohol, say, is aromatic, and you may think that it's easy to smell, but it has to be present in concentrations of at least 100 ppm. However, we can detect compounds such as octenone and β-damascone, both of which have distinctive aromas and are present in wine, in amounts of only 5 and 2, respectively, parts per *trillion*. In contrast, the vapor pressure of geologic minerals and almost all nutrient minerals is almost too small to measure—they have no smell. There are a few obscure geological exceptions—the arsenic sulfide mineral called arsenopyrite, for example, which to some has a slight whiff of garlic—but they are irrelevant to vineyards and wine aroma. A few metals show some tendency to sublimate—that is, change directly from solid to vapor—but they are so unstable as elements that they barely exist in nature. Moreover, they typically have unpleasant odors.

This lack of taste and odor is easily tested if you have access to a rock saw and some different kinds of rock. Smooth, freshly sawn surfaces will give a cool, tactile sensation on your tongue, but they will have no aroma or taste. No matter how much you lick and smell them, you won't be able to tell the rocks apart. (Surfaces that are not fresh are likely to taste of other things.) If you've ever noticed a geologist putting earth in his mouth, he won't be trying to taste it; he'll instead be chewing it to test the grain size. Particles finer than 0.002 millimeter, where silt gives way to clay, cease to feel gritty between the teeth. But there's no taste. Consequently, apart from a few halide, borate, and sulfide minerals, flavor is not a property mentioned in catalogues of mineral properties. In line with this, in none of the various textbooks and treatises on wine flavor are geologic materials mentioned. Tasting rocks and geologic minerals in wine has to be imaginary, in the mind.

The Taste Reminds Me of Stones

We all know, however, how helpful it can be when we are trying to describe certain perceptions in a wine to recourse to things geological. But they are metaphors, mental associations, and recollections of some past encounter involving rocks, not things actually present in vines and wines. We may even conjure what, say, a stone *should* taste like. For example, commentators often mention a flinty taste in a wine, perhaps most commonly in wines from Chablis and Sancerre (and, it has to be said, especially in the case of Sancerre if it's known the vines were growing in soils with pebbles of flint). But the Chablis soils have no flint and, conversely, those from, say, the Margaux district of the Médoc or the Martillac region of the Graves do have flint pebbles but yield wines that are not normally said to have a mineral flavor. Flint itself, like the other forms of silica (Chapter 3), is wholly without taste or smell, which is exactly why glass is made primarily of silica. So it's a curious thing to report a taste of flint (silica) as literally due to the vineyard ground if the wine has been stored in a glass bottle (silica) and is being tasted in a wine glass (also silica).

Many wine lovers, however, may at some time have struck together a couple of lumps of flint, and while it's tricky to get much of a spark, a distinctive smell is very noticeable. It's probably the extremely fine-grained nature of flint, together with its common trace impurities of sulfur and iron, that allows some vaporization to occur and hence produce a smell. But this is the case only if enough percussive energy is applied, which obviously makes all this irrelevant in any literal way to growing vines—the flinty taste or smell is metaphorical. And because "flinty" wines tend to be high in acid—sharp tasting—we might also be subconsciously recalling, to some extent, the well-known sharpness of broken flint edges.

An intriguing variation on this matter is the often mentioned "gunflint "aroma. It's odd that this aroma is mentioned so frequently in wine writings because probably few of us have experienced a gunflint smell, unless we play with antique firearms. For a long time now, the "flint" in guns has been synthetic and not geologic flint, which is also the case with "flint" lighters for gas stoves, barbecues, cigarettes, and the like. Most ways of reliably making sparks involve iron (or steel) because it has a property technically called **pyrophoricity**. Certain substances—the elements sodium, potassium, and calcium are examples—in the presence of oxygen are capable of spontaneously bursting into flames because they are extremely **pyrophoric**. This is why they don't occur as pure elements in nature. The phosphorus used in match heads needs the addition of a little heat, such as from the friction of striking a match, and a few metals like iron can be pulverized such that the surface area of a flying tiny particle is exposed to sufficient oxygen for it to ignite, making a spark. Iron readily makes sparks when struck on a tough substance like flint, but the familiar smell is burning iron, not flint.

Many of our mental associations suggest something geological, but in reality they are triggered by highly aromatic organic compounds. The air all around us is replete with bacteria, algae, molds, and the like, and at the Earth's surface they settle and rapidly film geologic surfaces. Some give rise to familiar odors. For example, the well-known smells of rocks and stones on a hot summer day, after a shower of rain, or upon being wetted at a river's edge are all due to the vaporization of certain organic oils (lipids, carotenoids, etc.) collectively termed **petrichor**.

An earthy smell in wine is similarly due to airborne organic compounds (such as some terpenes) common on vines and in wineries and are called **geosmin**. Some of them have astonishingly low sensory thresholds, down to a few parts per trillion. We may have smelled them if we watched soil being tilled, and we subconsciously recall the smell when describing a taste sensation in wine. Most metals lack any odor. Yet we often talk of a metallic smell, perhaps when we are handling coins or metal implements. The odor arises not from the metals themselves but through our having touched them, which causes a reaction between skin chemicals and the metal giving easily sensed aromatic compounds. In other words, while we conveniently label perceptions by referring to familiar inorganic materials, they are actually produced by aromatic compounds from the world of organic chemistry.

As a final example, wines from Chablis are sometimes said to be characterized by the iodine smell of the ocean. Some writers explicitly say that they can taste iodine in the wine and that it originated in the vineyard soils. Can it be coincidence that Chablis vineyards are well known for their fossilized seashells? Actually, like the gunflint example mentioned earlier, this comparison with iodine is curious because probably few people will have smelled it, let alone tasted it. Iodine can vaporize, though to nothing like the extent of many of the aromatic compounds in wine, but textbooks list the element as pungent, irritating—and toxic. (Domestic medicinal iodine—"tincture of iodine"—is dissolved in an organic solvent such as ethyl alcohol, and so the aroma for this iodine is quite different.) Iodine is a large ion and is generally unable to fit in the lattices of geologic minerals. It's the same with vegetation, unless it's a specially adapted form like seaweed. So we wouldn't expect much of it to be in vines. In line with this thinking, analyses show that the actual iodine content of the soils and vine leaves at Chablis is minuscule, and even less in the finished wine (around 4 ppb). Surprisingly perhaps, these iodine concentrations are less than in some nearby non-Chablis vineyards and less than half that of some New World Chardonnay wines that attract no mention of iodine. In other words, the perception of iodine also has to be a metaphor and unrelated in any direct way to the actual vineyard geology.

Science Begins to Show Some Connections

Despite all the doubts, queries, and negatives that I've just raised about connecting vineyard geology and wine flavor without substantiation, science has recently found some correlations broadly in line with popular ideas. Most of them link something about the rocks and soil with either certain chemical constituents of wine or make statistical correlations with sensory evaluations. However, the mechanisms underlying the links aren't clear, and none of the correlations are straightforward. As someone once said, 110% of statistics is misleading. And the majority of the correlations come down to the role of water in the vineyard and, to a lesser extent, the nutrient status of the soil. A pervasive difficulty is isolating which factor or factors are most significant in nature: is it necessarily the geology that's causing the effect? It's standard practice in laboratory experiments to hold all the relevant factors constant except for one, to explore how it behaves, and then, in turn, to evaluate each of the other parameters. But this is virtually impossible in a vineyard, with so many indeterminate and interacting variables.

For example, Grenache wines from coarser-grained, hence better-drained, soils in the Conca de Barberà DO, Tarragona Province, northeast Spain, show deeper color and different concentrations of flavor compounds compared with those from clay-based soils. *Ergo* drainage and water storage are crucial. However, the soils also differ in at least two parameters that have to do with fertility—organic matter and potassium content—which makes it hard to point to a governing parameter. Several studies in China, in vineyards on the alluvial plains of the Helam Mountain district, Ningxia Province, have correlated wine acidity, sugar, and phenols with a loose soil texture—in other words, good drainage. Apparently, because the three test vineyards were less than 20 kilometers apart, other parameters were *assumed* to be constant and therefore less important. However, similar research on Chardonnay wines from the Niagara Peninsula, Canada, while showing a role for soil drainage, concluded that the vineyard site and vintage year, that is, mesoclimatic factors and variations, were more influential.

The importance of water supply was demonstrated in studies at Geisenheim University in Germany on Riesling wines, where an undersupply increased certain flavor compounds (e.g., β-damascone) while lowering others (e.g., terpenes). Other work has shown that slight water deficits promote abscisic acid in grapes, which can lead to enhanced color and flavor. Of course, a restricted water supply to the vines, even in the absence of artificial manipulation, can arise from all sorts of different rocks and soils.

With regard to nutrient status, evidence exists that red wine color can be indirectly affected by the nitrogen uptake of a vine, through its influence on canopy growth and hence exposure to sunlight. In the La Côte vineyards of Vaud, Switzerland, for example, the sites on moraines have a lower humus content and hence less nitrogen availability than those located on nearby colluvium, and this leads to deeper colored wines with a more purple hue. Low nitrogen in grape juice can also affect the progress of fermentation because yeast, just like vines, requires nitrogen to thrive. But here not only is the role of geology highly indirect, but it's really the humus content that's relevant.

The La Côte work also noted that increased nitrogen resulted in wines with a lower pH, and acidity effects are another aspect for which there is much anecdotal evidence and a certain amount of data. The character of Pinot Noir wines from Oregon's Willamette Valley seems to vary with soil pH. A recent statistical study there related low pH soils to wines with rounded and complex flavors, whereas more alkaline soils produced wines with "brisk and less complex flavor," perhaps associated in some way with the soil pH controlling nutrient uptake.

There have been suggestions that a restricted potassium uptake can lead to a lower pH, which in turn yields wines of greater stability and color. This could explain why limestones are reported to yield "lively" and "refreshing" wines, attributes that presumably are due to acidity. The thinking is that an abundance of potassium, say from granitic soils, will react with a wine's tartaric acid to produce potassium bitartrate (a compound that involves hydrogen). This will tend to precipitate out, for instance, as the well-known crystals coating wine corks. This reduction of free hydrogen in the wine raises its pH and lowers its acidity. With limestone, however, because of its restricted potassium, the tartaric acidity of the wine remains intact.

But yet, besides the contradictory quotations mentioned earlier, there seems to be no empirical validation of the above idea about restricted potassium inducing lower acidity, and other research suggests an opposite effect: increased potassium uptake by vines *promoting* the production of both malic and tartaric acids. In line with this discrepancy, the experience in Germany, at least for Rieslings, is that limestone soils generally produce a wine of higher pH and lower perceived acidity than does sandstone or slate, which largely contradicts the thinking in France. The geological complexities that result from the thrust faults around St-Chinian, France, were illustrated in Chapter 7, but as a generalization limestone soils dominate in the south of the region and schist in the north. Yet, many commentators find more acid wines coming from the schist, contrasting with "fuller, more powerful" and "plush" wines coming from the limestone sites.

Perhaps the firmest scientific case for a connection between bedrock and wine flavor comes from the work of Ulrich Fischer and the Dienstleistungszentrum Ländlicher Raum (DLR) Rheinpfalz group, Germany, published in a 2011 American Chemical Society volume on the Authentication of Food and Wine. It concerns Riesling wines from the Pfalz and Mosel areas. Statistically comparing various taste sensations with water stress, growing degree days, temperature variation, solar radiation, and so on, they found that in some cases the best matches were with certain kinds of bedrock. In one study, for example, a panel of trained and tested judges first agreed on attributes of Riesling wines and then applied them to wines from a vineyard based on sandstone and another site nearby but on basalt.

Over two different vintages and in wines vinified separately by two different makers, they perceived systematic differences between wines from the two places, and the bedrock appeared to be an important factor. Similarly, wines from another site on sandstone were judged to be significantly different from wines from an adjacent vineyard based on rhyolite breccia.

The Rheinpfalz work nicely illustrates that the flavor profile of a wine can be influenced by the vineyard site, and it points to some involvement of bedrock. Noticeable are the corresponding statistical patterns of flavor attributes from the two different sandstone sites. They were consistently more vegetative than fruity, and they gave a "hard mouthfeel" and a "harsher acidity" than wines from the other kinds of bedrock.

The Rheinpfalz group also studied the aromatic compounds that may be responsible for the differing flavor profiles. For example, they found that wines from basalt typically had lemon/grapefruit aromas and a "smooth acidity" that involved compounds such as linalool and benzyl alcohol; graywacke gave lemon/grapefruit and green grass with "dominant acidity" and involved β-damascone; and limestone gave mango/passion fruit and peach/apricot with "smooth acidity," perhaps because of phenylethanol and the esters of propionic and acetic acids (think pineapples and nail polish!). The prominence of these characteristics did vary from year to year, and, of course, climatic and other factors are also involved. Even if we grasp the differences between "smooth," "harsh," and "dominant" acidity and accept that the statistical correlations are real, it remains unclear what mechanisms may be involved—what it is about the rocks that appears to be producing these effects.

A characteristic of some white wines, especially of some German Rieslings, is the "petrol" or "kerosene" aroma, which is now established as being due to a complex organic compound abbreviated as TDN. Wine lovers variously admire or dislike the taste, but apparently we can detect its presence in concentrations as low as 2 ppb. And there has been some excitement that the presence of TDN in a wine might be related to bedrock. Higher concentrations have been found in wines from slate soils, conceivably explaining its association with some classic Rieslings. Conversely, TDN appears to be suppressed by alkaline soils. Perhaps this is why it is rare in wines from, say, the limestones of Leithaberg in Austria. So finally there we have it: a clear link between bedrock and wine flavor!

But wait—again, it's nowhere near that simple. TDN also correlates with a whole range of properties, including wine age, bottle closure, cellar temperature, and, especially, exposure of the grapes to UV light. That's probably why a petrol flavor is more common in wines from the sunlit, elevated Eden Valley of Australia than from the less bright Adelaide Hills. And, of course, acidic and nitrogen-poor soils suppress canopy growth and leaf shading of the grapes. In other words, the bedrock may offer some correlations while by no means being a primary factor.

The Future: How Else Might Geology Affect Wine?

In Chapter 10, I speculated that some very particular and as yet inscrutable combination of the natural attributes of a site might underlie the greatness of certain terroirs and that this combination might carry through somehow to the nature of the finished wine. In addition, we saw in Chapter 9 how each mineral nutrient is needed within a certain optimal range. However, whether small differences *within* that span make any difference doesn't seem to have been explored. Most of the taste of wine comes from compounds produced during vinification—from so-called flavor precursors in the grape juice. Conceivably, because nutrients participate in reactions that produce these precursor compounds, variations within the optimal nutritional range might affect their concentrations and ultimately lead to differences in wine flavor. Metals such as zinc and iron are well known to influence how genetic information is translated into biological processes, so-called gene expression; perhaps variations in their amounts within the optimal range may have an effect.

The nutrients may possibly interact with each other, such that it is the balance between them that is significant rather than their actual concentrations. Iron, aluminum, magnesium, and copper have been shown to interact with the grape skin-derived anthocyanins that contribute to color and influence where it lies within the red-purple spectrum. In other words, slight variations in metal uptake can influence a wine's hue. How important such effects might be for flavor as opposed to appearance is pure speculation, but they could potentially strengthen the importance of soil nutrients for the finished wine.

In Chapter 10, I mentioned some of the new findings about distinctive communities of fungi and bacteria in different vineyards, what has been called a "microbial terroir." An intriguing new possibility is arising from such work, which might reveal a surprising role for geology in wine. The California Merlot work mentioned in Chapter 10 has linked the airborne bacteria with localized variations in the bacterial content of the soil, which seems to be acting as a kind of microbial reservoir. That is, the geology is to some extent influencing the airborne bacteria. But some new, highly precise observations are suggesting that certain soil bacteria live preferentially on particular minerals in the ground. A mineral's chemistry seems to be a factor, as is the physical nature of its surface. For example, at the ultramicroscopic scale, some bacteria seem to avoid minerals with smooth surfaces and prefer those minerals with cleavages (Chapter 2). It seems they like "sheltering" in the little steps where cleavages intersect with a mineral's surface. In other words, a connection might exist between the actual geologic minerals making the rocks and soils with the aerial microbiology. The research so far has been conducted in temperate forests rather than in vineyards, but there seems no reason in principle why a link shouldn't apply more generally.

It's wholly unclear what significance such bacteria might have for wine flavor, but here we have a glimpse of a possible new link between geology and wine, of how the dichotomy between anecdote and scientific understanding might be bridged. But all this is in the future. With regard to the situation at the moment, the reader can judge, in the light of all the matters just outlined, whether or not the prominence that's given to vineyard geology in today's populist writings is justified. Unless the scientific position changes, I can't really see that it is.

Of course, to many wine lovers all the above matters are transcended by the pure enjoyment of wine, a sensual appreciation that defies scientific analysis. Ultimately, our experience of drinking wine can no more be evaluated in terms of wine aroma and flavor than our response to a piece of music can be dissected by studying the score or measuring the audio frequencies. In addition, some enthusiasts imbue wine with a spiritual dimension, which, by definition, cannot be analyzed. But in this book we have been concerned with aspects of wine that can be addressed scientifically, so let's return to where we began, to the very start of this book. Let's remind ourselves that besides influencing growing vines, and all the technicalities we have outlined about tasting and wine flavor, geology has another role in vineyards, another relevance to wine lovers, one that doesn't affect wine directly but that can enrich our enjoyment of it. It's so fundamental that we often overlook it.

The Wonder of It All

A wine enthusiast shown untitled photographs of, say, the Champagne vineyards on the Montagne de Reims, the curves of the Douro slopes around Pinhão in Portugal, or the terraces on the hills making the Kaiserstuhl in Baden, Germany, would probably immediately recognize the places because they are so distinctive. Why so? Because the geology is different! As we saw in Chapter 8, it's ultimately the bedrock geology that determines the landscape and thus basically what a vineyard looks like. So as we contemplate the vineyards of the world, not only is the geology influencing the vine growth and, arguably, the character of the wines produced from them, but it's defining their very appearance. Geology is determining the scene. By using what we have learned in preceding chapters, we can view it with greater insight and thereby enhance our enjoyment of the wines it gives us.

Imagine the classic vistas as one looks down the Middle Mosel in Germany. Probably most striking is the river's tortuous course and the remarkable steepness of some of the vine-covered slopes. All this comes about because for a very long time, back in the Mesozoic era, this whole region was being worn down to a largely planar land surface, a flat floodplain across which the rivers lazily meandered. But then, about 800,000 years ago, it was geologically uplifted in the Mosel area by about 200 meters. The river was forced to start downcutting again, to carve the incised meanders we see today. Erosion at the outer bends is still cutting sideways into the bedrock, and this gives the cut bank and hence the steep vineyards—the celebrated Goldtröpfchen, for example, with the village of Piesport squeezed at its foot. On the opposite bank is the more sprawling community of Müstert, built on the point bar, the flat sediments deposited at the meander's inner arc.

The bedrock here is part of what is called the Rheinisches Schiefergebirge, the *Schiefer* indicating that the hills are dominated by slate, formed while the rocks were still deep below the land surface of the time, about 300 million years ago. And it may be the barren-looking slaty rubble in the vineyards that next catches your eye. Or it may be the handsome villages, again such as Piesport, with its roofs of the same dark gray Devonian slate as in the vineyards and the slender spire of the much-photographed church, also clad in the same local slate.

Imagine standing on Canada's Niagara Peninsula, say on the lake bench between Beamsville and Jordan, looking out across Lake Ontario, with Toronto hazy on the distant skyline. Below your feet, the gray-brown vineyard soil is derived from stony till modified by calcareous material slipping down from the abrupt hill behind you. That cliff, clothed in sugar maples, is the scarp face of the Niagara Escarpment, part of the cuesta that in Ontario runs roughly eastsoutheast–westnorthwest along the peninsula. Its orientation was determined about 20,000 years ago by glacial ice eroding in this direction, along a prominent set of joints in the bedrock dolomite. Some geologists have hypothesized that this influential joint direction is fundamentally related to stresses still operating in the North American tectonic plate you're standing on, still spreading northwestward from the mid-Atlantic Ridge divergent boundary.

But just here on the Beamsville bench, the joints developed locally in a northeast–southwest direction, which was also picked on by the eroding ice sheet and here gives an anomalous trend to the scarp. The resulting slightly more westerly aspect modifies sun exposure and focuses the prevailing southwesterly air flow from Lake Ontario, enhancing the lake effect which is so important in this part of the world. It gives the wineries located here their own individual mesoclimate. A secondary joint set at right angles to the scarp face has in places been dissolved away to give prominent fissures in the cliff, even narrow caves, which have attracted all kinds of local lore. One cave used to be famous for preserving ice long into the hot summers; another had a "fountain of youth," according to a local witch who claimed to be 300 years old. This legendary "cave spring" gives its name to a notable winery here.

Such geologically based features may not affect the wine in your glass much, but they may enhance how you think about it. I'm pointing here to a broader context, the bigger picture. And the background to a wine is clearly something of importance to wine lovers, as a glance at any wine magazine shows. It's why we pay premium prices for single vineyard wines, even though, logically, a skillfully blended fine wine should taste better.

Moreover, for me a feel for the geology not only informs my appreciation of vineyard vistas, but it alerts me to details as well: the slate being used as straining posts in a Muscadet vineyard; the chocolate-hued ironstone in old buildings (some roofed with slate imported from Wales) of the Barossa and Clare valleys; the contrasting pale gneisses and dark amphibolite in the walled terraces of the Wachau; the increased fissility of the bedrock dolomite at Flat Rock Winery, Ontario, hence its name; the local cream limestone polished for some of the worktops at the Ksara winery, Lebanon; the lapilli being used as a mulch in Lanzarote; the little ooliths in the walls of Chassagne-Montrachet, for centuries built from the Jurassic limestones in the quarries right behind the village.

As a final example of the big picture, visualize the vineyards of the Uco Valley in Argentina, newly fashionable and increasingly photographed. They stretch across the alluvial *brazos* of the broad flat valley, but two things may catch your eye, especially at the valley's northern end in the Tupungato department. First, there are the patches of boulders, not unusual in alluvial deposits such as these but some of them here reaching a remarkable 2 or more meters across. Their shapes vary, but all are well rounded, having been abraded and smoothed during their riverborne journey from high in the Andes. And not surprisingly, inasmuch as large parts of the Andes are composed of granite (and the eponymous andesite), most of the boulders are granitic.

So what *is* surprising is that the second striking feature, the valley's dramatic backdrop of snowcapped Andean mountains, is not here dominated by granite. On the skyline is the volcano Tupungatito, which spews out andesite and basalt. It erupted as recently as 1986 and is currently categorized as "restless." It's a reminder that although the ocean is over 200 kilometers away, deep, deep below the ground here, the tectonic plate that floors the Pacific is relentlessly grinding beneath the South American plate on which these Uco vineyards are sited.

There's a historical perspective as well. In 1835, a young English geological pioneer passed this way, having come across the Andes from Chile, assiduously noting everything he saw. That included the already established flourishing vineyards and other plentiful fields of fruit—he "bought water-melons nearly twice as large as a man's head, for a halfpenny apiece." One night he was bitten by a repulsive insect—and proceeded to keep it for several months as a kind of pet to study how long its blood meals would last. His geological observations were many and varied, and led to a string of geological discoveries, including the then radical idea that the Andes aren't ancient but geologically very young. And not least, he drew a marvelous geologic cross section across the Andes (in outline not greatly different from the modern equivalent) as he returned across the mountains from Mendoza to rejoin his ship in Valparaiso. The name of the ship? HMS *Beagle*. The young man (who more than once said his first love was geology) was Charles Darwin.

In conclusion, then, I believe that while a wine from the Uco Valley—or wherever—can be perfectly enjoyable in its own right, it becomes so much more than just a liquid in a glass when put into this kind of context. For me, such geological perspectives enrich the tasting experience and add immeasurably to my appreciation of the wonderful world of wine. It's not just drainage, nutrients, reflected sunshine, and the rest: vineyard geology also offers something bigger. And as for the matter of how it is that geology and all those multifarious other aspects of nature come together to create the wine we enjoy, well, that's the wonder of it all.

INDEX

Numbers in bold indicate a main entry; those in italics refer to a figure